CW01649307

Software Order Form

A Simple Introduction to Numerical Analysis 2: Interpolation and Approximation

Software forms an essential part of this Computer Illustrated Text. It provides worked examples, calculating power where necessary and is a great aid to understanding and learning.

Software is available for BBC and for IBM PC and compatibles on 51/4 disc at £25.50 (inc VAT). Site licenses are available for multiple use of software at £135.00 (inc VAT). Packs include master disc for copying, back-up disc and two books.

To receive your software complete this form and return with your remittance to: Adam Hilger, IOP Publishing Ltd, Techno House, Redcliffe Way, Bristol BS1 6NX, England Telephone: 0272 297481 Telex: 449149 INST P G

Telephone orders are accepted from customers paying by credit card

Please send me software for *A Simple Introduction to Numerical Analysis 2: Interpolation and Approximation* by R D Harding and D A Quinney on

☐ BBC disc ☐ IBM PC disc

☐ Site license BBC ☐ Site license IBM

I enclose my cheque for £_______ payable to IOP Publishing Ltd

Please charge to my Access/Visa/AmEx/Diners Club card number

☐☐☐☐☐☐☐☐☐☐☐☐☐☐☐☐ Expiry date _______

Name____________________ **Signature** ____________________

Address ____

____________________________ **Date** ________________

A Simple Introduction to Numerical Analysis

Volume 2 : Interpolation and Approximation

Other titles in the series

An Introduction to Groups
D Asche

Introduction to Statistics
A W Bowman and D R Robinson

Regression and Analysis of Variance
A W Bowman and D R Robinson

An Introduction to the Digital Analysis of Stationary Signals
I P Castro

Fourier Series and Transforms
R D Harding

A Simple Introduction to Numerical Analysis
R D Harding and D A Quinney

From Number Theory to Secret Codes
T Jackson

Electric Circuit Theory
B E Riches

Introduction to Probability
D R Robinson and A W Bowman

A Computer Illustrated Text

A SIMPLE INTRODUCTION TO NUMERICAL ANALYSIS
VOLUME 2 : INTERPOLATION AND APPROXIMATION

R D Harding

University of Cambridge

and

D A Quinney

University of Keele

Adam Hilger, Bristol and Philadelphia
ESM, Cambridge

British Library Cataloguing in Publication Data

Harding, Robert D.
A simple introduction to numerical analysis
Vol. 2 Interpolation and approximation
1. Numerical analysis
I. Title II. Quinney, Douglas, *1949-*
III. Series
519.4

ISBN 0-85274-154-5
ISBN 0-85274-156-1 IBM disc
ISBN 0-85274-157-X BBC 40/80 track disc
ISBN 0-85274-155-3 Network pack

Library of Congress Cataloging-in-Publication Data are available

Series Editor: **R D Harding**, University of Cambridge

Published by: IOP Publishing Ltd
Techno House, Redcliffe Way, Bristol BS1 6NX, England
242 Cherry Street, Philadelphia, PA 19106, USA
ESM, Duke Street, Wisbech, Cambs PE13 2AE, England

Published under the Adam Hilger/ESM imprint

Printed in Great Britain by JW Arrowsmith Ltd, Bristol

Contents

Preface

This book and disc is the second in a series of Computer Illustrated Texts (CITs) on topics in Numerical Analysis. The first, which was called *A Simple Introduction to Numerical Analysis*, covered topics such as the solution of algebraic equations, recurrence relations, linear equations, simple quadrature and the numerical solution of ordinary differential equations. In addition to the text each book is complemented by a disc of computer programs which are used as dynamic illustrations of the concepts discussed in the text and also remove much of the tedious calculation which is involved with repetitive processes. It is this integration of a standard text with computer software which makes a CIT. This second volume extends the material of *A Simple Introduction to Numerical Analysis* to consider topics in approximation. These topics include interpolation, i.e. fitting a smooth curve through specified points in order to determine non-tabular values, the approximation of functions by polynomials and rational functions, least squares fitting of data, numerical quadrature and the solution of boundary value problems involving differential equations using finite element techniques.

It is intended that the reader should have access to a suitable computer as he/she is reading the text and that at appropriate times the suggested programs are run. No detailed knowledge of computer programming is required though it is helpful if the reader is familiar with the keyboard and how to input standard functions. However, full details of how to use the package are given in Chapter 1.

This CIT is intended to be independent of *A Simple Introduction to Numerical Analysis*, however, at certain points in the text cross references are made to the original package. The reader will find it useful if this package is available but it is by no means essential.

RDH
DAQ
May 1988

Acknowledgments

The authors would like to thank Mike Bielby of Birmingham University, Malcolm Littler of Queensland Institute of Technology and Tony Croft of Leicester Polytechnic who have read previous drafts of the text and made numerous suggestions. They would also like to thank Margaret Downing who drew the diagrams.

The development of this CIT has been supported by the CIText group which has been financed by the joint UGC/Computer Board Computers in Teaching Initiative.

The text of this CIT was prepared by the authors using LaTeX and reproduced by IOP Publishing Ltd using camera-ready copy.

> Chapter 1

> Getting Started

> 1.1 Introduction

This book is the second in a series of Computer Illustrated Texts on Numerical Methods; the first was called *A Simple Introduction to Numerical Analysis.* A Computer Illustrated Text (CIT) consists of a software package which is integrated with a textbook. It is intended that the reader should, ideally, use the software frequently whilst working through the book, as the textbook and software components have been designed to complement and reinforce each other. However, it is also intended that the textbook component can be used without access to a computer, and that the software can be used on its own, for example, to generate demonstration programs or to solve problems beyond the scope of this introductory text. It is the combination of text and programs which gives an ideal teaching environment. Each CIT is self-contained in that all the programs which are required are either included in the text or supplied on the disc. However, the nature of the subject means that no short introductory textbook can include all the material required and so explicit references are given to other sources that may be needed. No ordering in terms of difficulty is implied between this CIT and the first, but some of the exercises in the text can be performed more easily if the disc associated with *A Simple Introduction to Numerical Analysis* is available. This first chapter describes how to run the programs, input data, store results and display graphics.

It is assumed that the reader has a basic knowledge of how to use a microcomputer but no knowledge of programming languages is needed, only an ability to enter data when required.

> 1.2 The software: an example

The software associated with this book was originally intended for the BBC microcomputer, but, other versions of the software will be available, and all instructions in the main part of the book are machine independent. However, in this section the features of the software will be introduced using the BBC version of the program **INTERP**. You may prefer to skip now to Chapter 2 as it is always possible to refer back to this chapter.

Users of machines other than the BBC microcomputer can usefully follow this section through, making only minor obvious variations. They will also find that their disc contains some introductory remarks which point out the most important of such variations.

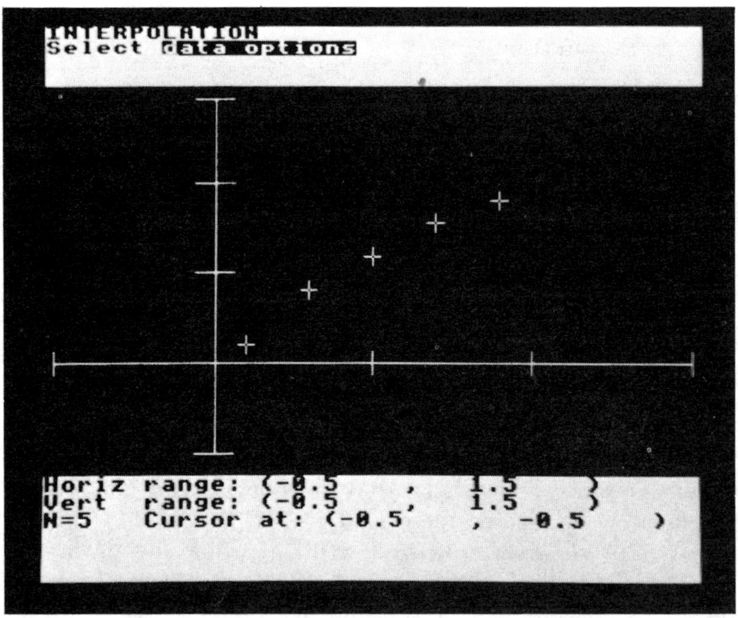

Figure 1.2.1 Initial screen setting of **INTERP**.

For BBC microcomputer users with the disc version of the software, loading of the program **INTERP** is accomplished by inserting the disc in drive 0, holding down SHIFT and then pressing and releasing BREAK. This should put up a menu on the screen giving a list of all the programs which are available. (On a network system it will be necessary to ask for advice regarding the appropriate loading instructions.) On some screens you may find that either the top line or the bottom line is obscured; this can be overcome by selecting the menu option **Screen adjustment** and then following the screen instructions to obtain the best setting. Once the menu is visible use the cursor keys to select the required program. Now select the program **INTERP**. The screen should clear and after a short delay it should look like figure 1.2.1. The screen has been split into three windows. The top window is used for dialogue, the bottom window contains current settings and the middle window is used either for graphic or tabular display. Initially, the option prompt should read

Select data options

where the **data option** is in inverse colours and a flashing cursor appears under the 'd' of data. The **data option** enables you to input, edit, plot and store data. Other options can be selected by pressing normal character keys: the SPACE bar is very convenient for this purpose. Alternatively, the key marked ↑ can be used to cycle forwards and the key marked ↓ to cycle backwards through the options available. An option is selected by pressing RETURN, but do not press RETURN just yet.

Whenever the program is loaded a default set of data is supplied and plotted; the five default points are shown in the display window by crosses. These values can be tabulated and a method for doing so will be described later. In the bottom window the current ranges of plotting are given together with the number of data points (N). It is also possible to input data using the cursor, the current position of which is also given in the bottom window. It is a uniform convention, used throughout the software, that all data which may be changed by the user will have its existing setting left unchanged if the user presses RETURN or ENTER. Where options need to be selected, the alternative may be viewed by pressing the SPACE bar repeatedly until the desired option is showing, then selected by pressing RETURN or ENTER.

The purpose of the program **INTERP** is to construct smooth curves through given data points. However, in this chapter we will only be interested in how data can be input, examined, edited, plotted and stored. Therefore, ensure that the **Select** option reads **data options** and then press RETURN to select this option. A new prompt should appear:

Data: from keys

Cycle through the options available at this point by pressing the SPACE bar repeatedly. If at any time you wish to quit press ESCAPE which will take you back to the following prompt:

Action: Continue

Try it and see. Since we wish to continue, accept the default setting by pressing RETURN; the program should then advance to the **Select data options** prompt as above. Press RETURN again and we arrive back at the **Data** prompt.

Select the option **Data: from keys** by pressing RETURN. Next we have the option of clearing any existing data points. For the moment let us just add to the existing points by using the default setting, i.e. by pressing RETURN. Additional points can either be input in tabular or graphical form. Pressing RETURN again will select the option

Data input format: tabular

and display the current points. A flashing cursor should now appear at the bottom of the table and instructions concerning the input of data are given in the top window. Try adding the point (0.95, 0.95) by typing in each coordinate followed by RETURN. If you make a mistake you can delete characters using the DELETE key before correcting them but each coordinate is terminated by RETURN. To quit press Q. This process will return the program to the option **Data: from keys** after sorting the tabular points into ascending order. Now press RETURN to accept the current option of supplying data from the keyboard and then press RETURN again to retain the existing points. The program now advances to the option **Data input format: tabular**. Press the SPACE bar to change this to **Data input format: graphical** and then press RETURN. All the current data points are displayed on the default graphical settings. These can be changed at this point using the **Graph data** option. Cycle through the options available, **superimpose**, **clear - same axes** and **clear - new axes**. For example, the last of these options allows the user to set up different sets of axes. When this option is shown press RETURN. The flashing cursor now appears in the bottom window and it is possible to alter the range of each axis in turn. For the moment accept the default settings by pressing RETURN each time the flashing cursor moves to a new position in the bottom window. When all the required values have been input the axes will be redrawn.

We will next investigate the graphical editing options. Firstly, select the option **Add point**. It is now possible to add additional points either by (i) specifying the x and y coordinates or (ii) using the cursor. For (i) first press 'X' and a flashing cursor will appear in the bottom window as the x coordinate of the cursor. Try inputting a value 0.05 followed by RETURN. Now press 'Y' and input a y coordinate of 0.0 followed by RETURN. The flashing cursor in the middle window is now at the position (0.05, 0.0). In order to add this point to the table press RETURN again, this will be confirmed by an increase in N in the bottom window. To try out method (ii), first select the **Add point** option again by pressing RETURN. Additional points can now be added by moving the cursor with the arrow keys. The increment by which the cursor moves is controlled by the '<' (increase) or '>' (decrease) keys and is displayed in the top window. Note that the two methods can be combined; for example, you may set the x coordinate and then adjust it using the cursor control keys. The x and y values are open to change up to the moment when you press RETURN.

The other graphical editing options are **Change point**, **Delete point** and **Quit editor**. The **Change point** option first asks you to select a point by using the cursor control keys; at this stage each keystroke makes the cursor jump to the next point in ascending order (right or up arrow), or descending order (left or down arrow). When the cursor flashes on the point you wish to change press RETURN, and the situation is exactly as it was with the **Add point** option. The **Delete point** option also begins by asking you to select a point; once you have selected it you are asked to confirm that you wish to delete the current point. The **Quit editor** option takes the program back to the option level from which the editor was invoked.

At any stage of data input it is possible to view the current set of data points by escaping from the current option and selecting the option

Data: tabulate

An option to send any tabulation to a connected printer is available which reminds you to ensure that the printer is switched on. The option

Data: plot &/or edit

is available to display or modify any existing data set.

It is possible to store data in a file using the option

Data: store on file

and retrieve suitable data using

Data: from file

Wherever possible data files created by one program can be used as input for any other when it is appropriate, but if an inappropriate data file is requested then the program will display a suitable error message. Five simple data files are provided on the program disc[1], as follows:

DATA
SIN
SPLINE
RAMP
FE

When loading a file you will need to type the whole file name at every occasion, however, you need not distinguish between upper and lower case. You will not be able to store any data on the program disc which is supplied because it is write protected and so you will need an unprotected disc of your own. You can give the data files any name which is consistent with the rules of the filing system you are using. Try loading the file **DATA**, examining it, editing it and then storing the result.

> 1.3 The software: general remarks

The previous section demonstrated how to supply simple numerical data and also the use of cyclical prompts. Although the actual prompts and options will vary from program to program the repeated use of the RETURN key will always accept the current default setting and consequently the first time a program is run results from the default parameters should agree with those given in the text. Throughout the programs the use of ESCAPE will always return the program to the **Action** prompt from which the program can be restarted or terminated as required. The ultimate panic button on the BBC model B is the BREAK key; this should only be pressed in emergencies. For MS-DOS versions BREAK and CTRL/C have the same effect as ESC.

An alternative to numeric input is to supply data by means of expressions. For example, at any point where the exponential number 'e' is required it is possible to type the string EXP(1). However, if the string supplied cannot be evaluated then a message such as 'bad input' or 'error evaluating function' will be given. Such an error will not necessarily be fatal to the program: a suitable error message will be given and usually the user will be invited to type in the item again. Some

[1] With BBC discs data files have a prefix D.; MS-DOS files will have a .DAT extension.

errors cannot be detected until later in the calculation: these will also give rise to a suitable message, but it will act like an ESC in that the program will jump to the **Action** prompt.

> 1.4 Screen dumps

It is frequently useful to obtain a permanent copy of the screen display for later examination. Programs running under MS-DOS may use the key labelled **Print Screen**, or an equivalent such as **Prnt Sc**, to obtain hard copies of the screen, provided that a suitable printer has been attached and has been selected by using the GRAPHICS command (consult your manual). For BBC microcomputer users, there is, unfortunately, no universal standard for printers or the software to drive them. Therefore, rather than supply a program which gives a direct screen-to-printer dump we have provided a facility for copying the screen image to memory, usually a disc file. The dump file can then be used to recreate the screen image and users can then provide their own software to make a hardcopy on any printer available. The details of this process will depend on the hardware available and the system being used. For the BBC microcomputer to which a disc drive is connected it is possible to obtain a screen dump as follows. Whenever a **dump** option is shown press RETURN at which point the top window should show

Dump to filename:

Make sure that the disc in the drive is not write-protected and then type in a suitable file name, which must satisfy the usual BBC filename conventions, followed by RETURN.

If an Epson printer, or any other device which is software compatible, is to be used then the screen dump can be copied to the printer using the program **PRNTDMP** which is available from the main menu. Selecting this program gives a self-explanatory dialogue for the user to follow.

One final advantage to this method of obtaining screen dumps is that it is possible to keep a permanent copy of screen images and recall them to the screen for future demonstration. For BBC users this can be carried out from within the program or else in intermediate command mode. In the latter case if the file DMPF were the name given to a dumpfile created in mode 4 then it can be recalled to the screen, but not printed, using the command

MODE 4:*LOAD DMPF

> 1.5 Final remarks

If you want to use numerical methods in more complicated programs, as opposed to learning about them, and you wish to write programs in BBC BASIC, then *A Mathematical Toolkit: Numerical Routines with Applications in Engineering, Mathematics and the Sciences* (Harding 1986) contains subroutines which can be built into your own programs. For graphical output, *Graphs and Charts on the BBC Microcomputer* (Harding 1982) contains a library of subroutines for this purpose. It is intended that in the near future the CIT series will contain a *Mathematical and Graphical Toolkit* for MS-DOS machines.

> Chapter 2

> Interpolation

'Interpolation - the art of reading between the lines.'

> 2.1 Polynomial interpolation

We shall begin our discussion of interpolation by considering the following problem.

Example 2.1.1
In an experiment the yield of a chemical species is recorded as a function of the concentration of another reagent which is present. As it is impossible to vary the concentration of the reagent continuously, only a finite set of different values can be tabulated. Let us assume that the following results have been obtained.

Table 2.1.1 Data for example 2.1.1.

Concentration of reagent	Yield
0.10	0.09956
0.30	0.39646
0.50	0.58813
0.70	0.77210
0.90	0.89608

We can now ask questions like:

1. Is it possible to estimate the yield when the reagent has a concentration of 0.51?

2. How accurate is such an approximation going to be?

3. Are the results which are produced sensitive to perturbations or errors in the tabular values? Clearly, we would like small changes in the data to produce small changes in the results. If this is the case then the problem is said to be *well-conditioned.*

4. Is it possible to determine a reagent concentration which gives a prescribed yield, for example, find a concentration such that the yield is 0.5? (This is the inverse problem to (1)).

The problem of attempting to estimate intermediate values from a given table is called *interpolation* and is the basis for the following sections.

In order to provide a background for understanding the methods described in this chapter we shall begin by considering the problem of obtaining non-tabular values in elementary ways. We assume that we are given a table of values (x_i, y_i), $i = 0, \ldots, n$, and that we need to determine a non-tabular point (x, y). For example, let us suppose that we wish to find an approximate value which corresponds to $x = 0.51$ in table 2.1.1. Perhaps the simplest way is to join together successive points as shown in figure 2.1.1. In particular, between $x = 0.5$ and $x = 0.7$ we have the straight line shown in figure 2.1.2 and we can find the coordinates of the point C by linear interpolation.

From figure 2.1.2 the point C has as its y coordinate $y(0.5) + CD$. However, by similar triangles,

$$\frac{CD}{AD} = \frac{BE}{AE}$$

therefore,

$$CD = AD\frac{BE}{AE} = 0.01\frac{(0.77210 - 0.58813)}{(0.7 - 0.5)} = 0.00920,$$

to five significant figures. (Only five significant figures are quoted here since the original tabular values are only given to this accuracy.) Hence, $y(0.51)$ is approximately $0.58813 + 0.00920 = 0.59733$.

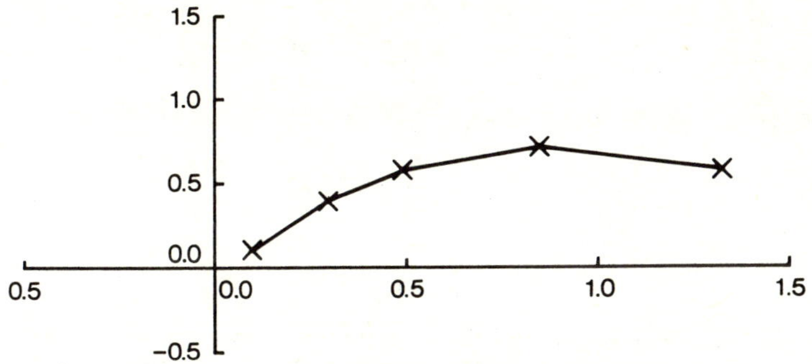

Figure 2.1.1 A simple linear interpolant for table 2.1.1.

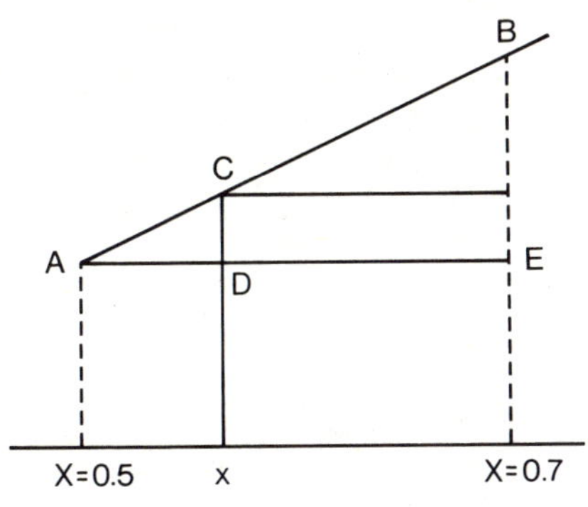

Figure 2.1.2 Linear interpolation between two points.

Instead of using the similar triangle argument above we could have determined an explicit expression for the equation of the straight line joining the points A and B in figure 2.1.2. This line will be of the form

$$y = c_0 + c_1 x,$$

and passes through the points (0.5, 0.58813) and (0.7, 0.77210). Therefore,

$$0.58813 = c_0 + 0.5c_1$$

and

$$0.77210 = c_0 + 0.7c_1.$$

Eliminating c_0 gives $c_1 = (0.77210 - 0.58813)/(0.7 - 0.5) = 0.91985$, which is the gradient of the line. The intercept c_0 is then found to be at 0.12821, hence the equation of the line joining the points A and B is

$$y = 0.12821 + 0.91985x,$$

from which $y(0.51) = 0.59733$ as before. Clearly, this very naive approach is simple to use but ignores most of the information given in the table. Furthermore, it fits an unnatural function to the data supplied since the fitted function has a strange derived function. Between each pair of consecutive data points the fitted function is linear and therefore has constant slope. The derivative of the piecewise linear function shown in figure 2.1.1 will therefore be piecewise constant and look like the function shown in figure 2.1.3. In order to overcome this problem we could use higher order functions rather than linear ones. For example, through any three points we can fit a unique parabola.

Exercise 2.1.1 (Analytical/Computational) Find the parabola through the points (0.3, 0.39646), (0.5, 0.58813) and (0.7, 0.77210). (Hint: the general parabolic function is $y = a + bx + cx^2$, and the three coefficients can be determined by substituting the three points into this equation to produce three linear equations which must be solved to find values for a, b and c.)

If the idea of fitting a polynomial through a number of points is unfamiliar, then it is important to work through exercise 2.1.1. In particular, check that the resulting quadratic function takes the correct values at $x = 0.3$, $x = 0.5$ and $x = 0.7$. It is not difficult to extend this process to work for arbitrary polynomials, and it is to this problem that we next turn our attention.

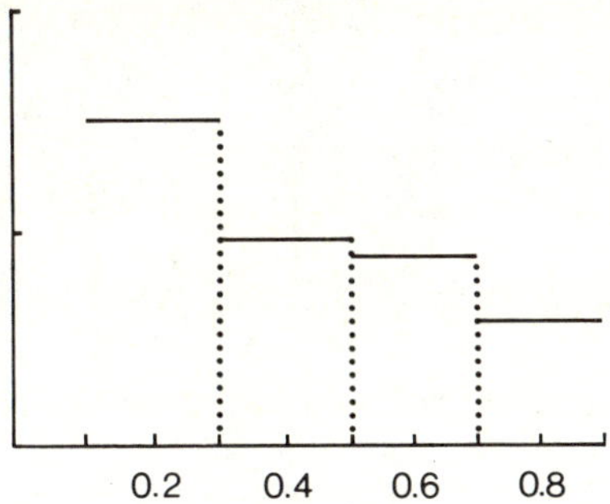

Figure 2.1.3 The derived function for figure 2.1.1.

A polynomial of degree n has $n+1$ coefficients and so we can fit a unique polynomial of degree n through $n+1$ distinct tabular points, x_i, $i = 0, \ldots, n$. Let us assume that we are given a table of values (x_i, y_i), $i = 0, \ldots, n$, and try to fit such a polynomial. Let P_n be the *interpolating polynomial*

$$P_n(x) = c_0 + c_1 x + c_2 x^2 + \ldots + c_n x^n.$$

Now we select the coefficients c_j, $j = 0, \ldots, n$, so that the equations

$$P_n(x_i) = y_i, \qquad i = 0, \ldots, n, \tag{2.1.1}$$

are satisfied. This gives a system of $n+1$ *linear* equations in the $n+1$ unknown coefficients c_j, $j = 0, \ldots, n$. These linear equations may then be solved using, for example, Gaussian elimination with partial pivoting. (See Harding and Quinney (1986).)

Example 2.1.2

For the five values in table 2.1.1 we can fit a unique polynomial of degree 4, i.e.

$$P_4(x) = c_0 + c_1 x + c_2 x^2 + c_3 x^3 + c_4 x^4.$$

Evaluating this polynomial at the five data points in table 2.1.1 gives the following system of linear equations:

$$\begin{pmatrix} 1 & 0.1 & 0.1^2 & 0.1^3 & 0.1^4 \\ 1 & 0.3 & 0.3^2 & 0.3^3 & 0.3^4 \\ 1 & 0.5 & 0.5^2 & 0.5^3 & 0.5^4 \\ 1 & 0.7 & 0.7^2 & 0.7^3 & 0.7^4 \\ 1 & 0.9 & 0.9^2 & 0.9^3 & 0.9^4 \end{pmatrix} \begin{pmatrix} c_0 \\ c_1 \\ c_2 \\ c_3 \\ c_4 \end{pmatrix} = \begin{pmatrix} 0.09956 \\ 0.39646 \\ 0.58813 \\ 0.77210 \\ 0.89608 \end{pmatrix}. \tag{2.1.2}$$

Using Gaussian elimination, the solution of this system of equations is

$$[-0.159796, 3.164656, -6.499406, 8.274375, -3.901563]^T$$

which produces the interpolating quartic

$$P_4(x) = -0.159796 + 3.164656x - 6.499406x^2 + 8.274375x^3 - 3.901563x^4.$$

(You can easily check these results using the program **GAUSS** from Harding and Quinney (1986).) Evaluating P_4 at the given points produces the values in table 2.1.2. (Later we shall provide an alternative simpler method.)

Table 2.1.2 Polynomial interpolation applied to table 2.1.1.

x	y	$P_4(x)$	$P_4(x) - y$
0.1	0.09956	0.0995598	2.4×10^{-7}
0.3	0.39646	0.3964597	2.8×10^{-7}
0.5	0.58813	0.5881297	3.1×10^{-7}
0.7	0.77210	0.7720996	3.9×10^{-7}
0.9	0.89608	0.8960794	5.7×10^{-7}

Notice that the values in the third column of table 2.1.2 are displayed to seven decimal places even though the data are only given to five. This is to emphasize the fact that whenever this process is carried out it will be prone to numerical errors. For this example, if we round the results in column 3 to five decimal places we recover the initial data. Whether this is always the case will be examined later. The function obtained, $P_4(x)$, is shown in figure 2.1.4. Notice that between the tabulated points the interpolating polynomial is quite smooth and so we are justified in evaluating it at non-tabular points. For example, we can determine a

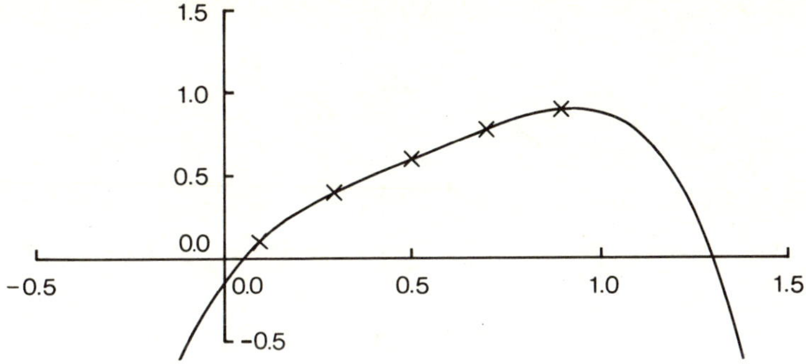

Figure 2.1.4 Quartic polynomial for table 2.1.1.

value for $y(0.51)$ by $P_4(0.51) = 0.59733892$. However, this is not always the case as the next example will demonstrate. Outside the range of the tabular points it would be unwise to use the interpolating polynomial to extrapolate the given values. This can be seen by examining the curve in figure 2.1.4 for values of x in excess of 0.9 where one might expect a monotonic increasing function.

Example 2.1.3
The interpolating polynomial which passes through the points $(-2, e^{-4})$, $(-1, e^{-1})$, $(0, 1)$, $(1, e^{-1})$, and $(2, e^{-4})$ is shown in figure 2.1.5. These values are obtained from the function $\exp(-x^2)$ which is always positive but as figure 2.1.5 shows the same is not true of the interpolating polynomial for values of x in the intervals $(-2, -1)$ and $(1, 2)$! Furthermore, as the number of points increases the interpolating polynomial behaves less like the function $\exp(-x^2)$. This is called *Runge's problem* and will be discussed in more detail in the next chapter. (See also exercises 2.1.4 and 3.4.4.)

Exercise 2.1.2 (Computational) The program **INTERP** is an interactive program which can be used to construct interpolating polynomials through a set of data points. The values in table 2.1.1 are set up as the default tabulated points. Therefore, load the program **INTERP** and when the screen shows

Select data options

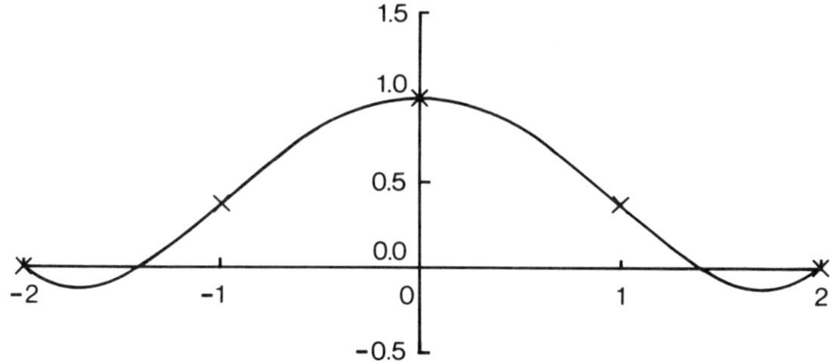

Figure 2.1.5 Runge's problem: see example 2.1.3.

press the space bar which will change the option to **Select method**. Press RETURN and the option prompt will change to

Method: Construct interpolating polyn.

You can cycle through the various methods by repeatedly pressing any character key except RETURN; now select the option above by pressing RETURN, which will calculate the required interpolating polynomial passing through the points in table 2.1.1. First of all, you are allowed the option of selecting the graphics parameters which are used to plot the resulting polynomial. For the moment select the option

Graph data: superimpose

The interpolating polynomial is now drawn in the graphics window and the actual polynomial is given in the bottom window. (The coefficients are limited to five significant figures by the space available to display them.) We can now evaluate the required interpolating polynomial using the option

Option: Evaluate interpolant

Entering the value for $x = 0.51$ produces the same value as in example 2.1.2. All computation is carried out to full accuracy; only the coefficients are displayed to five significant figures.

Exercise 2.1.3 (Computational) Use the program **INTERP** and any four values from table 2.1.1 to determine a cubic interpolant and compare the estimate for $y(0.51)$ with the value determined above. This can be carried out by quitting the current option by pressing Q and then selecting the option **Edit data**. Press RETURN in order to accept the current plotting parameters and then cycle through the edit options available by pressing any key but RETURN. (Recall that you can cycle backwards using the key marked ↓.) When the current option is **Delete point** press RETURN and use the cursor keys to select the point you wish to remove and then press RETURN again. Cycle through the options now available until the option is **Quit editor** and then press RETURN. Once again cycle through the options available until the current option becomes **Select method** and press RETURN yet again. Accepting the option **Method: Construct interpolating polyn.**, the interpolating polynomial through the remaining points will be determined and plotted. The effect of deleting another point can be examined as follows. Cycle until the option **Edit data** is current and then press RETURN twice. This will take you into the data editor; the current edit option should be **Add point**. If this is the case press RETURN, if not cycle until it is and then press RETURN. The cursor in the graphics window should now be flashing at the point which was previously deleted; this point can be added back to the table simply by pressing RETURN. Other points can then be deleted to consider the effect of fewer points in the table. How does the selection of the four points influence the interpolated value?

Exercise 2.1.4 (Computational) Use the program **INTERP** to obtain five equally spaced points in the interval $(-5, 5)$ from $y = \exp(-x^2)$ and then determine a suitable interpolating polynomial (see example 2.1.3). You will need to take the default prompt **Select data option** and then **Data: from keys** remembering to clear the existing data before using the **Data input format: tabular** option. For example, the point $(-5, \exp(-25))$ can be input as the tabular values $X(1) = -5$ and $Y(1)$ =EXP(−25). When you have constructed the interpolating polynomial through the points you have selected, try adding other points and see how the results compare. What happens as the number of points increases/decreases? Why is it better to have an even number of points? Investigate what happens with non-equally spaced points.

In general, given $n + 1$ tabular points, (x_i, y_i), $i = 0, \ldots, n$, we can proceed as above and determine the coefficients of the interpolating polynomial $P_n(x)$ by solving a system of linear equations given by equation (2.1.1), or equivalently

$$\mathbf{Ac} = \mathbf{y}, \tag{2.1.3}$$

where $\mathbf{A}$ is an $n+1$ by $n+1$ matrix and $\mathbf{c}$ and $\mathbf{y}$ are vectors of dimension $n+1$. The columns and rows of $\mathbf{A}$ are numbered 0 to n and the components of $\mathbf{A}$ are $a_{ij} = x_i^j$, $i = 0, \ldots, n$, $j = 0, \ldots, n$. Similarly $y_i = y(x_i)$, $i = 0, \ldots, n$, and c_i are the coefficients of the interpolating polynomial which is required. (See equations (2.1.2) for a simple example with $n = 4$.) This system of equations will have a unique solution provided the matrix $\mathbf{A}$ is non-singular, therefore, we examine the possibility of it being singular or almost singular. In order to do this we consider the determinant of $\mathbf{A}$; this matrix is called a *Vandermonde matrix* and it is possible to show that it has determinant

$$\prod_{i=1,\ldots,n,\ j<i} (x_i - x_j).$$

Exercise 2.1.5 (Analytical) Show that the determinant of the Vandermonde matrix of order 3, given by

$$\begin{pmatrix} 1 & x_0 & x_0^2 \\ 1 & x_1 & x_1^2 \\ 1 & x_2 & x_2^2 \end{pmatrix},$$

is $(x_1 - x_0)(x_2 - x_0)(x_2 - x_1)$.

The importance of the above expression for the determinant of the Vandermonde matrix is that if the tabulation points are close together then the determinant may be small, i.e. the equations are nearly singular, and hence the solution may be inaccurate. This type of inaccuracy is usually called *inherent ill-conditioning* and is a fundamental part of the problem, i.e. independent of the method of solution. In order to demonstrate the problems which may arise when determining interpolating polynomials in such circumstances let us consider the following example.

Example 2.1.4 (Computational)
Let us begin by finding the quartic polynomial which passes through the points in table 2.1.3. Using the program **INTERP** produces the interpolating polynomial shown in figure 2.1.6. In order to consider how sensitive this problem is let us perturb the y value of the third entry in table 2.1.3 from 1.05 to 0.95 and see how this affects the computed solution. The resulting quartic polynomial is shown in figure 2.1.7. Compared with the original interpolation quartic it is clear that this problem is inherently ill-conditioned.

Table 2.1.3

x	y
0.00	2.00
0.95	1.00
1.00	1.05
1.05	1.00
2.00	0.00

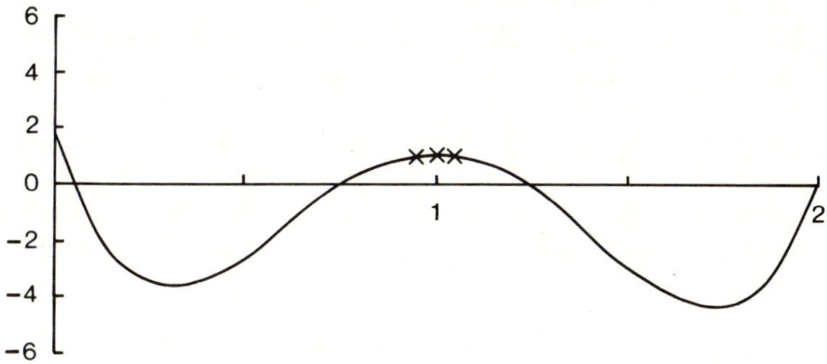

Figure 2.1.6 The quartic interpolant for table 2.1.3.

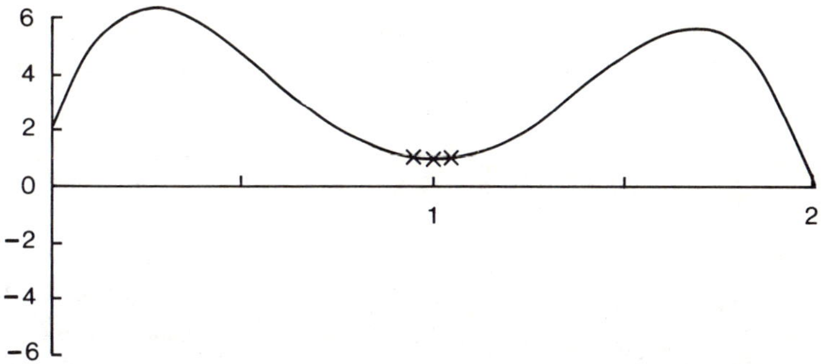

Figure 2.1.7 The effect of perturbing entry 3 in table 2.1.3.

Example 2.1.4 gave an illustration of a problem which was inherently ill-conditioned, i.e. a small change in the tabular values leads to a much larger change in the interpolating polynomial. This was caused by several of the tabular values being close together, resulting in a system of linear equations which were nearly singular. We can also investigate the possibility of errors being introduced in the interpolating polynomial by the method we have selected to solve this problem. Such perturbations are called *induced errors.* Ideally we would like small changes in the tabular values to produce small changes in the interpolating polynomial and if this is the case then the problem is said to be well conditioned. In order to demonstrate this idea consider the following example.

Example 2.1.5 (Computational)
Applying the program **INTERP** to construct an interpolating polynomial through the points in table 2.1.1 produces the polynomial

$$P_4(x) = -0.159796 + 3.164656x - 6.499406x^2 + 8.274375x^3 - 3.901563x^4.$$

Now let us round the entries in table 2.1.1 to four decimal places and then construct a polynomial through the resulting points. This gives the polynomial

$$-0.1597 + 3.1652x - 6.5044x^2 + 8.2833x^3 - 3.9063x^4.$$

The maximum relative change in any coefficient is less than 0.12% and so this problem is relatively well conditioned.

Exercise 2.1.6 (Computational) Use the program **INTERP** to construct the interpolating polynomial through the tabular values in the file **SIN**. Investigate the sensitivity of the data provided by examining the effect of perturbing one or more of the values given. Is this problem inherently well conditioned? Is it susceptible to induced ill-conditioning?

> 2.2 Lagrangian interpolation

In order to construct a polynomial of degree n through $n + 1$ distinct points using equation (2.1.1) it is necessary to solve a system of linear equations. We will now derive an alternative way of producing such a polynomial which does not require the solution of a system of possibly ill-conditioned equations. As there is only one polynomial of degree n which passes through $n+1$ points we would obtain the same polynomial as that obtained from equation (2.1.1) if exact arithmetic were used. (See exercise 2.2.4.)

Let us begin by assuming that we have been given a set of data points, (x_i, y_i), $i = 0, \ldots, n$, and then examine the function

$$(x - x_0)(x - x_1)\ldots(x - x_{i-1})(x - x_{i+1})\ldots(x - x_n).$$

Notice that there is no factor $(x - x_i)$. This function can also be written as the product

$$\prod_{j \neq i}(x - x_j).$$

and is a polynomial of degree n which vanishes at each data point with the exception $x = x_i$. We now define the *Lagrangian polynomial functions*, $L_i(x)$, $i = 0, ..., n$, as

$$L_i(x) = \frac{\prod_{j \neq i}(x - x_j)}{\prod_{j \neq i}(x_i - x_j)}, \qquad i = 0, \ldots, n. \tag{2.2.1}$$

We see that each of these polynomials has degree n and that $L_i(x_k) = 0$ if $k \neq i$ but is equal to 1 when $k = i$. Now consider the function

$$P(x) = \sum_{i=0}^{i=n} y_i L_i(x). \tag{2.2.2}$$

Since $L_i(x)$ is a polynomial of degree n for each i, so is $P(x)$ and by virtue of the way in which each of the functions L_i are defined $P(x)$ will take the values y_i when $x = x_i$, $i = 0, \ldots, n$. Therefore, $P(x)$ and the interpolating polynomial $P_n(x)$ are identical since they are both polynomials of degree n which agree at $n + 1$ points.

Example 2.2.1 (Analytical)

Let us consider the data points shown in table 2.1.1. The Lagrangian polynomial L_0 is given as follows:

$$\begin{aligned} L_0(x) &= \frac{(x - 0.3)(x - 0.5)(x - 0.7)(x - 0.9)}{(0.1 - 0.3)(0.1 - 0.5)(0.1 - 0.7)(0.1 - 0.9)} \\ &= 26.04167x^4 - 62.5x^3 + 53.64583x^2 - 19.375x + 2.460937. \end{aligned}$$

We stress that in practice an explicit expression of such a polynomial would not be determined; it is easier to evaluate it in the form of equation (2.2.2). You should check that $L_0(x_0) = 1$, i.e. $L_0(0.1) = 1$, and that $L_0(x_i) = 0$, $i = 1, 2, 3, 4$, i.e. $L_0(0.3) = L_0(0.5) = L_0(0.7) = L_0(0.9) = 0$. However, because of the form of equation (2.2.2) it is beter to consider $y_i L_i(x)$, for example, $y_0 L_0(x)$ is plotted in figure 2.2.1.

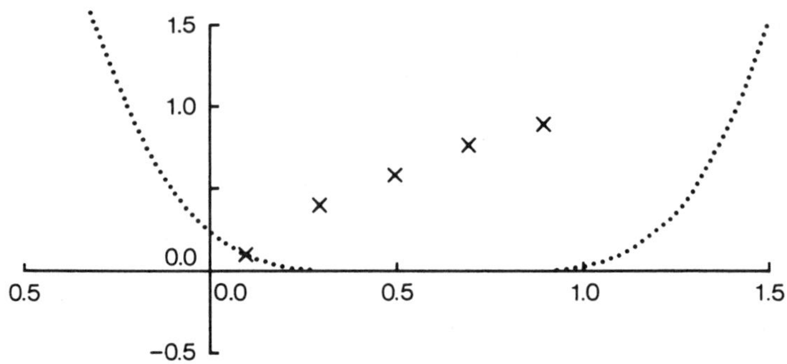

Figure 2.2.1 The Lagrangian polynomial $y_0L_0(x)$.

Similarly, we can construct the polynomials $L_1(x)$, $L_2(x)$, $L_3(x)$ and $L_4(x)$. The required interpolating quartic polynomial is then given by

$$P_4(x) = 0.09956L_0(x) + 0.39646L_1(x) + 0.58813L_2(x) + 0.77210L_3(x) + 0.89608L_4(x).$$

To evaluate $P_4(0.51)$ we firstly find $L_0(0.51) = 0.004052$ from equation (2.2.1) and then, similarly, we have that $L_1(0.51) = -0.03165$, $L_2(0.51) = 0.99688$, $L_3(0.51) = 0.03498$ and $L_4(0.51) = -0.00426$, hence $P_4(0.51) = 0.59734$ as before.

Exercise 2.2.1 (Computational) The program **INTERP** may be used to construct a suitable interpolating polynomial through the points given in table 2.1.1. Load **INTERP**, cycle through the options available with the **Select** prompt until it reads **method** and then press RETURN. Cycle again until the **Method** prompt is **Lagrangian polynomials** and press RETURN again. Now use the default **Graph** option which is **superimpose** and a flashing cursor will appear in the graphics window. Different points can be selected using the cursor keys and then pressing RETURN once again. For example, the Lagrangian polynomial $L_0(x)$ can be drawn by simply pressing RETURN when the flashing cursor is positioned over the first tabular point. However, in order to make use of the form of equation (2.2.2) it is more convenient to use the functions $y_0L_0(x)$ and it is this function which is plotted in the graphics window and displayed in the bottom window. In a similar way the other four

Lagrangian polynomials can be drawn.

Exercise 2.2.2 (Computational) Use the program **INTERP** to construct the Lagrangian polynomials through the data in file **DATA**. Sum these polynomials as in equation (2.2.2) and show that the result is equivalent to the polynomial given by direct construction of the interpolating polynomial.

Exercise 2.2.3 (Computational) Use the program **INTERP** to determine suitable third- and fourth-order polynomial interpolants for the data given in table 2.1.1. (Hint: use the editor to delete points one at a time.)

Exercise 2.2.4 (Analytical) Show that there is a unique polynomial of degree n which passes through the $(n+1)$ points $(x_i, y_i), i = 0, \ldots, n$. (Hint: consider two polynomials $p_n(x) = a_0 + a_1 x + \ldots + a_n x^n$ and $q_n(x) = b_0 + b_1 x + \ldots + b_n x^n$, with $a_i \neq b_i$ for at least one i, each of which interpolates the points given. Now consider the non-trivial polynomial $p_n(x) - q_n(x)$ at the points x_i and hence show that $a_i = b_i$, $i = 0, \ldots, n$.)

Exercise 2.2.5 (Analytical) Hermite interpolation. Lagrangian interpolation is based upon tabular values (x_i, y_i), $i = 0, \ldots, n$; if in addition the gradients y_i' are supplied then higher order polynomials can be fitted to the data. Show that it is possible to fit a polynomial of degree $2n+1$ in the form

$$P_{2n+1}(x) = \sum_{i=0}^{i=n} \alpha_i(x) y_i + \sum_{i=0}^{i=n} \beta_i(x) y_i'$$

where

$$\begin{aligned} \alpha_i(x) &= [1 - 2L_i'(x_i)(x - x_i)] L_i(x)^2, \\ \beta_i(x) &= (x - x_i) L_i(x)^2. \end{aligned}$$

The result is called a Hermite interpolation polynomial.

> 2.3 Divided differences

As another alternative to the construction of an interpolating polynomial through the points (x_i, y_i), $i = 0, \ldots, n$, using Lagrangian polynomials we can determine the interpolating polynomial of degree n as follows.

If we define the function $p_0(x) = y_0$ then, trivially, $p_0(x_0) = y_0$. Now consider the linear polynomial

$$p_1(x) = y_0 + a_1(x - x_0),$$

which satisfies $p_1(x_0) = y_0$ and then select $a_1 = (y_1 - y_0)/(x_1 - x_0)$ so that

$$p_1(x_1) = y_0 + a_1(x_1 - x_0) = y_1.$$

Next define

$$p_2(x) = a_0 + a_1(x - x_0) + a_2(x - x_0)(x - x_1),$$

then since $a_0 = y_0$ and $a_1 = (y_1 - y_0)/(x_1 - x_0)$ we have that $p_2(x_0) = y_0$ and $p_2(x_1) = y_1$. Furthermore,

$$p_2(x) = p_1(x) + a_2(x - x_0)(x - x_1).$$

We still have a_2 at our disposal, but if

$$a_2 = \frac{(p_2(x_2) - p_1(x_2))}{(x_2 - x_0)(x_2 - x_1)}$$

then $p_2(x_2) = y_2$, i.e. $p_2(x)$ passes through (x_0, y_0), (x_1, y_1) and (x_2, y_2). We can repeat this process by defining

$$p_j(x) = p_{j-1}(x) + a_j(x - x_0)(x - x_1)(x - x_2)...(x - x_{j-1}),\ j = 1, \dots, n,$$

and then selecting a_j so that $p_j(x_j) = y_j$. The resulting polynomial, $p_n(x)$, passes through (x_i, y_i), $i = 0, \dots, n$, and is identical with that obtained by using the Lagrangian polynomials.

Example 2.3.1 (Analytical)
The interpolating polynomial through the first three values in table 2.1.1 may be constructed as follows. Using the above method gives

$$\begin{aligned} p_0(x) &= y_0 = 0.09956, \\ p_1(x) &= p_0(x) + a_1(x - x_0) = 0.09956 + a_1(x - 0.1). \end{aligned}$$

Select a_1 so that $p_1(x_1) = 0.39646$, i.e. $a_1 = \frac{(0.39646 - 0.09956)}{(0.3 - 0.1)}$ which gives

$$p_1(x) = 0.09956 + 1.4845(x - 0.1).$$

Next we define $p_2(x)$ by

$$p_2(x) = p_1(x) + a_2(x - 0.1)(x - 0.3)$$

and select a_2 so that $p_2(x_2) = y_2$, i.e. $p_2(0.5) = 0.58813$,

$$\begin{aligned} a_2 &= \frac{(p_2(x_2) - p_1(x_2))}{(x_2 - x_0)(x_2 - x_1)} \\ &= \frac{(0.58813 - 0.69336)}{(0.5 - 0.1)(0.5 - 0.3)} = -1.315375. \end{aligned}$$

This gives the polynomial

$$p_2(x) = 0.09956 + 1.4845(x - 0.1) - 1.315375(x - 0.1)(x - 0.3)$$

which passes through the first three entries in table 2.1.1.

The process outline above can be automated in the following way. Given the table of values (x_i, y_i), $i = 0, \ldots, n$, we define the *first divided differences* associated with these values by

$$f[x_i, x_{i+1}] = \frac{(y_{i+1} - y_i)}{(x_{i+1} - x_i)}.$$

Next we compute *second divided differences* as

$$f[x_i, x_{i+1}, x_{i+2}] = \frac{(f[x_{i+1}, x_{i+2}] - f[x_i, x_{i+1}])}{(x_{i+2} - x_i)}$$

and so on. When the $(k - 1)$th divided differences have been obtained, i.e.

$$f[x_i, x_{i+1}, \ldots, x_{i+k-1}] \text{ and } f[x_{i+1}, x_{i+2}, \ldots, x_{i+k}]$$

we define the kth divided difference as

$$f[x_i, x_{i+1}, \ldots, x_{i+k}] = \frac{(f[x_{i+1}, \ldots, x_{i+k}] - f[x_i, \ldots, x_{i+k-1}])}{(x_{i+k} - x_i)}.$$

These differences can be shown schematically as in table 2.3.1.

We can then show that the polynomial $p_n(x)$, given by

$$\begin{aligned} p_n(x) &= y_0 + f[x_0, x_1](x - x_0) + f[x_0, x_1, x_2](x - x_0)(x - x_1) \\ &\quad + \ldots + f[x_0, x_1, \ldots, x_n](x - x_0)(x - x_1) \ldots (x - x_{n-1}), \end{aligned}$$

passes through the points (x_i, y_i), $i = 0, \ldots, n$, and is therefore identical to the interpolating polynomial $P_n(x)$. (See exercise 2.2.4.)

Table 2.3.1 Table of divided differences.

x	y	First divided difference	Second divided difference	Third divided difference
x_0	y_0			
		$f[x_0,x_1]=\frac{(y_1-y_0)}{x_1-x_0}$		
x_1	y_1		$f[x_0,x_1,x_2]=\frac{f[x_1,x_2]-f[x_0,x_1]}{x_2-x_0}$	
		$f[x_1,x_2]=\frac{(y_2-y_1)}{x_2-x_1}$		$f[x_0,x_1,x_2,x_3]$
x_2	y_2		$f[x_1,x_2,x_3]=\frac{f[x_2,x_3]-f[x_1,x_2]}{x_3-x_1}$	
		$f[x_2,x_3]=\frac{(y_3-y_2)}{x_3-x_2}$		
x_3	y_3			

Exercise 2.3.1 (Analytical) Show that the above expression is the interpolating polynomial through the points (x_i, y_i), $i = 0, \ldots, n$.

Exercise 2.3.2 (Analytical) Show that $f[x_1, x_2, \ldots, x_k]$ is invariant with respect to all permutations of the tabular points $x_1, x_2, \ldots, x_k$.

Table 2.3.2 Tableau of divided differences for table 2.1.1.

x	y	First difference	Second difference	Third difference	Fourth difference
0.1	**0.09956**				
		1.48450			
0.3	0.39646		**−1.31538**		
		0.95835		**2.031875**	
0.5	0.58813		−0.09625		**−3.9015625**
		0.91985		−1.089375	
0.7	0.77210		−0.74988		
		0.61990			
0.9	0.89608				

Example 2.3.2 (Calculator)
To demonstrate the use of divided differences we construct the interpolating polynomial through the five points in table 2.1.1. Firstly, we obtain the table of divided differences given by table 2.3.2, as illustrated

in table 2.3.1. The required quartic polynomial is then given by

$$\begin{aligned} P_4(x) &= 0.09956 + 1.48450(x-0.1) - 1.31538(x-0.1)(x-0.3) \\ &\quad +2.031875(x-0.1)(x-0.3)(x-0.5) \\ &\quad -3.9015625(x-0.1)(x-0.3)(x-0.5)(x-0.7), \end{aligned}$$

where the coefficients are given by the first entry in each column of the tableau of divided differences. Expanding this expression gives the polynomial

$$P_4(x) = -0.159796 + 3.164656x - 6.499406x^2 + 8.274375x^3 - 3.901563x^4,$$

which agrees with the interpolating polynomial through these points as computed previously.

Exercise 2.3.3 (Analytical) Use the method given in section 2.1 to determine an interpolating polynomial of degree 2 which passes through the points $(0,3)$, $(1,6)$ and $(2,11)$. Obtain the corresponding Lagrangian polynomials L_0, L_1 and L_2 and hence show that the polynomial

$$3L_0(x) + 6L_1(x) + 11L_2(x)$$

is identical with the interpolating polynomial which was derived using equation (2.1.1). Construct a table of divided differences for these data points and hence construct a quadratic function which passes through the same points. Confirm that all three methods give the same polynomial.

Exercise 2.3.4 (Programming) Write a simple program to read in a set of tabular points and construct a table of divided differences. Test your program on the data given in exercise 2.3.3. Extend the program to produce polynomial interpolants as described in example 2.3.1.

> 2.4 Neville's algorithm

Although the previous sections demonstrated how it was possible to construct a suitable interpolating polynomial explicitly it is clear that a considerable amount of effort is required in order to interpolate from the values given in a table. Furthermore, neither the method outlined in section 2.1 nor the Lagrangian method gives any indication of the order of polynomial required. What we seek is some means of using a subset of the tabulated points, m say, to construct a suitable interpolant and then by adding an additional point construct a higher order polynomial. In this way if we can show that the interpolants produced using subsets of m points and $m+1$ points agree then there is no point in investigating higher order polynomials. Neville's algorithm does precisely this.

Let us suppose we select a subset of m points from a table of values, (x_0, y_0), (x_1, y_1), ..., (x_{m-1}, y_{m-1}). Next we construct the linear functions

$$\begin{aligned}
p_{01}(x) &= \frac{(x - x_0)y_1 - (x - x_1)y_0}{(x_1 - x_0)}, \\
p_{12}(x) &= \frac{(x - x_1)y_2 - (x - x_2)y_1}{(x_2 - x_1)}, \\
&\vdots \\
p_{m-2,m-1}(x) &= \frac{(x - x_{m-2})y_{m-1} - (x - x_{m-1})y_{m-2}}{(x_{m-1} - x_{m-2})}.
\end{aligned}$$

Notice that $p_{j,j+1}(x)$ passes through the points (x_j, y_j) and (x_{j+1}, y_{j+1}), $j = 0, \ldots, m-1$. (These terms are sometimes called *linear cross means.*) Next we construct the linear cross means between successive linear terms, e.g. between $p_{01}(x)$ and $p_{12}(x)$, and then evaluate the quadratic function

$$\frac{(x - x_0)p_{12}(x) - (x - x_2)p_{01}(x)}{(x_2 - x_0)}.$$

This function passes through (x_0, y_0), (x_1, y_1) and (x_2, y_2) and is denoted by $p_{02}(x)$. In this way we can recursively define a sequence of polynomials of increasing order by

$$p_{jk}(x) = \frac{(x - x_j)p_{j+1,k}(x) - (x - x_k)p_{j,k-1}(x)}{x_k - x_j}, k > j, k = 1, .., m. \tag{2.4.1}$$

This polynomial is of degree $(k-j)$ and passes through the points (x_i, y_i) for $i = j, \ldots, k$, and is therefore equivalent to the Lagrangian polynomial of the same degree (see exercise 2.2.4).

Exercise 2.4.1 (Analytical) Show that the quadratic polynomial $p_{03}(x)$ satisfies the following: $p_{03}(x_0) = y_0$, $p_{03}(x_1) = y_1$, $p_{03}(x_2) = y_2$.

In practice if we are given a set of data and require an estimate at a non-tabular point x^* then, rather than evaluating the general polynomial, we only evaluate (2.4.1) at the point x^*. Furthermore, it is usual to take the tabular points beginning at the point closest to x^* and taking subsequent points in increasing distance from x^* so that x^* is near the middle of the values used.

Example 2.4.1 (Computational)
Use Neville's algorithm to find an estimate for $y(0.51)$ in table 2.1.1 on page 9. Load **INTERP** and cycle on the **Select method** prompt to select **Neville's Algorithm**. We shall use the option **Direct** so press RETURN again, and then enter the value $x = 0.51$. The values in table 2.1.1 are set as default values and should appear in a window on the right of the screen. We can now select the entries from this table in order of closeness to the point $x = 0.51$. The tabular value $x = 0.5$ is the closest one to $x = 0.51$, so that is the first one we wish to select. To do so, use the up/down cursor keys to position the inverse video block over the required value and then press RETURN. Another inverse video block will now appear under the heading 'x, y selected', and you can copy your chosen values into this just by pressing RETURN. The next tabular values needed are those at $x = 0.7$, which should be selected and copied in a similar way. At this point type Q to stop selecting values and the program will then display a tableau based on the two values $x = 0.5$ and $x = 0.7$. Further points can be added by typing Q and selecting the **Select/edit data** option. Add further values until you obtain the tableau shown in table 2.4.1. Notice that some editing of the selected values is allowed (should it be needed). For example, you can delete an entry (the entries below then move up one place), or copy one of the data points on top of another already selected entry which then replaces that entry. (Only three columns of the table of cross means are visible on the screen at any one time; other columns can be viewed using the left/right cursor keys.) Notice that by taking the tabulation points in such a way that the values $|x^* - x_i|$ are in increasing order of size the top entry in each column gives the interpolated value required. When the entries in successive columns are sufficiently close we can terminate the process. Additional values can be added to the bottom of the table in order to improve the current interpolated value. In this example the second and third entries agree to three decimal places and so it is sensible to quote the result as $y(0.51) \approx 0.598$. (In fact the value of p_{03} is much better than this; from section 2.1, $P_4(0.51) = 0.59734$.)

Exercise 2.4.2 (Computational) Use the program **INTERP** with the option **Neville's algorithm** to estimate $y(0.51)$ by adding the final entry to table 2.4.1 from table 2.1.1.

Table 2.4.1 Neville's algorithm applied to table 2.1.1.

x_i	$(0.51 - x_i)$	y_i	Linear	Quadratic	Cubic
0.5	+0.01	0.58813			
			$p_{01} = 0.59733$		
0.7	−0.19	0.77210		$p_{02} = 0.59751$	
			$p_{12} = 0.59367$		$p_{03} = 0.59795$
0.3	+0.21	0.39646		$p_{13} = 0.6149$	
			$p_{23} = 0.57133$		
0.9	−0.35	0.89608			

Unfortunately, Neville's algorithm does not always work as well as shown in example 2.4.1. If the tabulated data points are sufficiently smooth then all is well but consider the following example.

Example 2.4.2 (Computational)
Table 2.4.2 illustrates the application of the program **INTERP**, with option **Neville's algorithm**, to estimate $y(0.15)$, from the values given. The result is poor because the tabulated values are obtained from the function $y(x) = 1/x$ which changes very rapidly near $x = 0$. (The correct value of $y(0.15) = 6.666666$, illustrating the difficulties which may appear if the data points vary greatly.)

Table 2.4.2 Results produced by Neville's algorithm.

x	y	Linear	Quadratic	Cubic	Quartic
0.1	10.000000				
		7.500000			
0.2	5.000000		7.083333		
		5.833333		6.927083	
0.3	3.333333		6.145833		6.848958
		4.583333		6.302083	
0.4	2.500000		5.208333		
		3.750000			
0.5	2.000000				

Exercise 2.4.3 (Computational) Use the program **INTERP** to retrieve data from file **SIN** and use Neville's algorithm to estimate $y(\frac{1}{3}\pi)$. The

data in this file is obtained from the function $y = \sin x$. Insert additional points and investigate how the accuracy of the interpolated point depends on the spacing and number of points.

> 2.4.1 Inverse interpolation

Thus far we have concentrated upon the problem of interpolating non-tabular values. In this section we will look briefly at the problem of inverse interpolation which was mentioned on page 10. The simplest solution to this problem is to reverse the roles played by the dependent and independent variables; for example, we simply rewrite table 2.1.1 on page 9 as table 2.4.3. This technique must be used with care because if there are two x values with the same y value then we will not get a single-valued interpolating function. This is equivalent to obtaining a singular matrix for equation (2.1.1). Furthermore, we need the original tabular values to be monotonic otherwise the order of the points will change when the roles of dependent and independent variables are interchanged.

Table 2.4.3 Interchanging variables in table 2.1.1.

Yield, y	Concentration of reagent, x
0.09956	0.10
0.39646	0.30
0.58813	0.50
0.77210	0.70
0.89608	0.90

Exercise 2.4.4 (Computational) Use interpolation applied to table 2.1.1 on page 9 to determine $x(0.4)$ using **Neville's algorithm** and the option **Inverse**. (Notice that applying inverse interpolation to this table is equivalent to applying interpolation to table 2.4.3. The program **INTERP** retains the original order of each (x, y) pair on the screen although their roles are reversed.) Check the result by using the option **Construct interpolating polyn.** and evaluating it at the point you have determined. How would you account for any discrepancy?

The procedure of interchanging the dependent and independent variable can be very successful but must be used with care. It is clear that interpolation works best when the x_i values are not too close and the y_i

values do not change too quickly. If this is the case then we can expect a reasonably accurate interpolation, but if these conditions are not satisfied then inverse interpolation may be difficult. However, in order to illustrate the usefulness of inverse interpolation consider the following simple problem.

Table 2.4.4 Inverse interpolation applied to $x^2 - 3x + 1 = 0$.

x	y					
0.4	−0.040					
		0.38261				
0.3	0.190		0.38189			
		0.38636		0.38195		
0.5	−0.250		0.38261		0.38196	
		0.39130		0.38213		0.38196
0.2	0.440		0.38392		0.38202	
		0.36296		0.38258		
0.1	0.710		0.37721			
		0.34483				
0.0	1.000					

Example 2.4.3 (Computational)
Use inverse interpolation to determine the solutions of the quadratic equation

$$x^2 - 3x + 1 = 0.$$

Tabulating this function between $x = 0$ and $x = 4$ shows that the solutions lie in the intervals $(0, 0.5)$ and $(2.5, 3)$. Now load the program **INTERP** and enter values for $f(x) = x^2 - 3x + 1$ for values of x between 0 and 0.5. Using **Neville's algortihm** and the option **Inverse** produces the results in table 2.4.4. Notice that even though the roles of the dependent variable x and the independent variable y have been interchanged they are displayed in the same order on the screen. From this table the required solution is at $x = 0.38196$. (The solutions of the given quadratic are at $x = \frac{1}{2}(3 \pm \sqrt{5})$, the smallest of which is 0.381966011.)

Exercise 2.4.5 (Computational) Find the larger root of the quadratic equation given in example 2.4.3 by tabulating the function $f(x)$ in the interval $(2.5, 3)$ and then using inverse interpolation.

Exercise 2.4.6 (Computational) Load the data from file **SIN** and use inverse interpolation to determine a value of x such that $y = 0.5$. The data in this file are obtained from the function $y = \sin x$; investigate the effect of additional points and different orderings.

> 2.5 Spline interpolation

Example 2.1.2 showed how it was possible to fit a polynomial of degree 4 through five points, but the resulting function was quite oscillatory. We will now investigate how it is possible to fit lower order interpolants which are globally 'smoother'. For example, through each pair of points (x_i, y_i) and (x_{i+1}, y_{i+1}) there is a unique straight line, but such a function is unsatisfactory since it will in general have kinks at each data point. (See figure 2.5.1.) Mathematically speaking, the interpolating function will not be differentiable at the points x_i, $i = 1, \ldots, n-1$. As an alternative, we could fit a quadratic function through (x_i, y_i) and (x_{i+1}, y_{i+1}). This problem does not have a unique solution, but we can take advantage of this non-uniqueness to select the quadratic curves whose derivatives match at each of the data points.

Figure 2.5.1 Local linear interpolation.

Let us suppose that we attempt to fit the quadratic function

$$Q_i(x) = a_i + b_i x + c_i x^2, \qquad i = 1, \ldots, n, \tag{2.5.1}$$

in the interval (x_{i-1}, x_i); then as there are n intervals we shall have $3n$ coefficients to find. First of all, we require that the resulting functions interpolate the points (x_i, y_i), $i = 0, \ldots, n$, i.e.

$$Q_i(x_{i-1}) = y_{i-1} \text{ and } Q_i(x_i) = y_i, \qquad i = 1, \ldots, n, \tag{2.5.2}$$

which gives $2n$ equations. (This condition will ensure that the resulting functions are continuous at the internal points $x_1, \ldots, x_{n-1}$.) Furthermore, at each internal point, x_i, we require that the derivatives of Q_i and Q_{i+1} agree, i.e.

$$Q_i'(x_i) = Q_{i+1}'(x_i), \qquad i = 1, \ldots, n-1. \tag{2.5.3}$$

Therefore, we have $3n-1$ equations to determine the $3n$ coefficients a_i, b_i, c_i, $i = 1, \ldots, n$. It would appear that we can select any one coefficient arbitrarily. Combining together the resulting quadratic functions gives a *quadratic spline*.

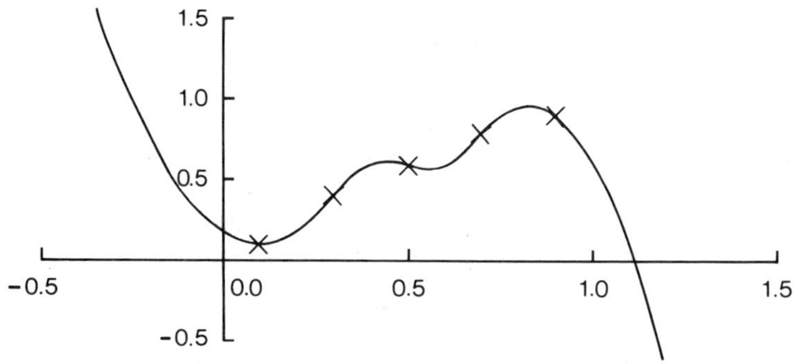

Figure 2.5.2 A quadratic spline through the data in table 2.1.1.

The amount of effort in determining a quadratic spline can be reduced by making use of the following procedure. Let us assume that in the interval (x_{i-1}, x_i) the required spline component is $Q_i(x)$, as shown in figure 2.5.3. We now assume that at x_{i-1} the slopes of both $Q_{i-1}(x)$ and $Q_i(x)$ are d_{i-1} and similarly at x_i, $Q_i'(x) = Q_{i+1}'(x) = d_i$. Since the derivative of a quadratic function is linear, $Q_i'(x)$ is the linear cross mean of d_i and d_{i-1}, i.e. we have that

$$Q_i'(x) = \frac{(x - x_{i-1})d_i - (x - x_i)d_{i-1}}{x_i - x_{i-1}},$$

therefore, $Q_i'(x_i) = Q_{i+1}'(x_i) = d_i$, which will ensure that the conditions (2.5.3) are satisfied. Next we integrate this expression to get

$$Q_i(x) = \frac{(x - x_{i-1})^2 d_i - (x - x_i)^2 d_{i-1}}{2(x_i - x_{i-1})} + c_i,$$

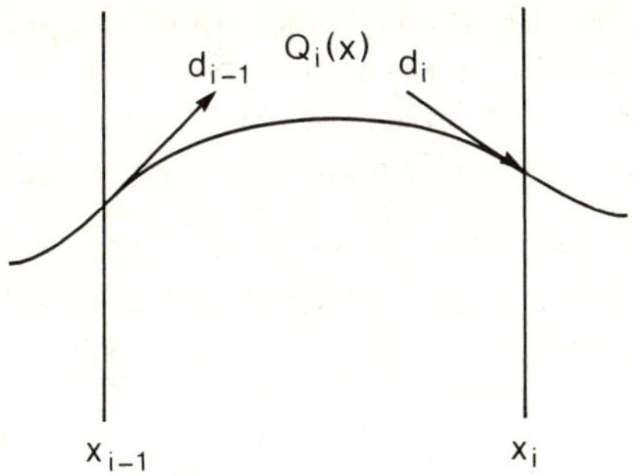

Figure 2.5.3 A quadratic spline component $Q_i(x)$ in (x_{i-1}, x_i).

where c_i is a constant produced by the integration. Finally, we select the constants c_i to ensure that the combined spline is continuous across each mesh node $x = x_i$, $i = 1, \ldots, n-1$. (This ensures that equations (2.5.2) are satisfied.) Notice that

$$Q_i(x_i) = \tfrac{1}{2}(x_i - x_{i-1})d_i + c_i = y_i$$

and

$$Q_i(x_{i-1}) = \tfrac{1}{2}(x_{i-1} - x_i)d_{i-1} + c_i = y_{i-1}.$$

We then eliminate c_i between these equations to give

$$d_i = \frac{2(y_{i-1} - y_i)}{(x_{i-1} - x_i)} - d_{i-1}. \tag{2.5.4}$$

This is a first-order recurrence relation and so all the required values, $d_1, d_2, \ldots$ can be found once d_0 is specified. However, in general we will not be able to specify both d_0 and d_n. (See Harding and Quinney (1986).)

Example 2.5.1 (Computational)
Load **INTERP** and when the prompt shows **Select method** press RETURN. Now change the option to **Method: Quadratic spline** and then press RETURN again. Use the graph option to set up suitable plotting ranges and then you will be requested to supply a value for the slope of the spline at its left end, i.e. the value d_0. For the moment accept the default value $d_0 = 1.4845$ and then observe the corresponding quadratic spline being plotted. (See figure 2.5.4.) Alternatively, we could take

$d_0 = 0$ which gives the quadratic spline shown in figure 2.5.2. The value $d_0 = (y_1 - y_0)/(x_1 - x_0)$ is obtained by using a finite difference approximation for the slope at x_0 which explains why a better result is produced. (This is the default value which the program adopts.)

Exercise 2.5.1 (Computational) Use the program **INTERP** to determine a quadratic spline through the data points in table 2.1.1. Investigate the effect of changing the value of d_0. (The quadratic spline for the default choice is given by figure 2.5.4.)

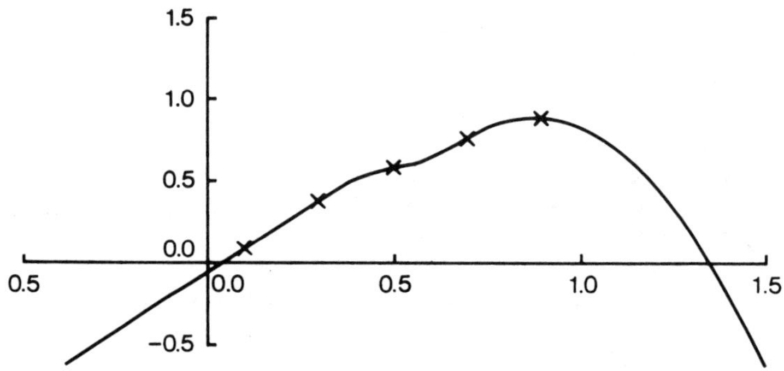

Figure 2.5.4 Quadratic spline for table 2.1.1 with $d_0 = 1.4845$.

From figure 2.5.4 we see that a quadratic spline and its derivative are continuous at each node. However, the curvature is not smooth since

$$\lim_{x \to x_1} Q_0''(x) = \frac{(d_1 - d_0)}{(x_1 - x_0)}$$

but

$$\lim_{x \to x_1} Q_1''(x) = \frac{(d_2 - d_1)}{(x_2 - x_1)}$$

and these may not be the same. A further problem with quadratic splines is that they tend to produce oscillatory functions as is also demonstrated in figure 2.5.2. In order to ensure that both the first and second derivatives are continuous we need to use a higher order local interpolant in each sub-interval. One of the most popular is to use cubic functions in each sub-interval and then attempt to select the coefficients so that

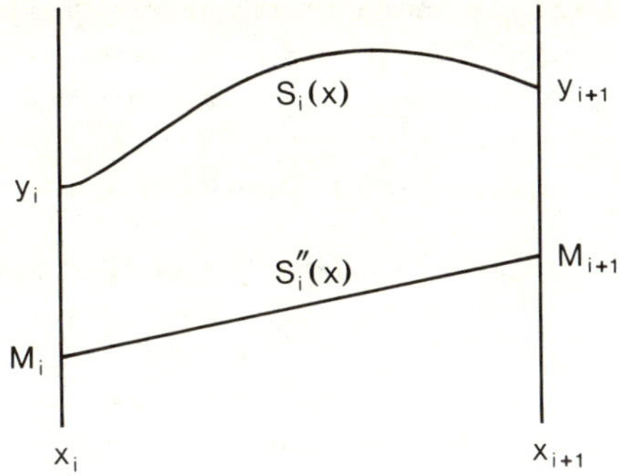

Figure 2.5.5 Cubic polynomial through (x_i, y_i) and (x_{i+1}, y_{i+1}).

both the slope and the curvature are continuous across internal tabular points.

Let us suppose that we are given a table of data in the form of data points (x_i, y_i), $i = 0, \ldots, n$. In the interval (x_i, x_{i+1}) we use the cubic polynomial

$$S_i(x) = a_i + b_i x + c_i x^2 + d_i x^3. \tag{2.5.5}$$

In each sub-interval there are four coefficients to be determined, and since there are n such intervals this will give a total of $4n$ unknown values. Clearly,

$$S_i(x_i) = y_i \quad \text{and} \quad S_i(x_{i+1}) = y_{i+1},\ i = 0, \ldots, n-1,$$

which gives $2n$ conditions. Similarly, at each internal point we require

$$S'_{i-1}(x_i) = S'_i(x_i),\ i = 1, \ldots, n-1,$$

and

$$S''_{i-1}(x_i) = S''_i(x_i),\ i = 1, \ldots, n-1,$$

which gives a further $2n-2$ conditions (see figure 2.5.5). It would appear that we can specify any two values arbitrarily. Once all $4n$ parameters are determined we shall have a unique cubic in each subinterval. This collection of functions is called a *cubic spline*. Even for a moderate number of tabulation points this procedure will result in a large system of linear equations but, fortunately, the total amount of effort can be considerably reduced in a manner analogous to that used for quadratic splines.

Notice that in the interval (x_i, x_{i+1}) we have that

$$S_i''(x) = 2c_i + 6d_i x. \tag{2.5.6}$$

Let us assume that the second derivative of S_i at x_i is given by some value M_i, then

$$\begin{aligned} S_i''(x_i) &= 2c_i + 6d_i x_i &&= M_i, \\ S_i''(x_{i+1}) &= 2c_i + 6d_i x_{i+1} &&= M_{i+1}. \end{aligned}$$

Solving these equations for c_i and d_i gives

$$6d_i = \frac{M_{i+1} - M_i}{x_{i+1} - x_i}, \qquad 2c_i = \frac{M_i x_{i+1} - M_{i+1} x_i}{x_{i+1} - x_i}.$$

Therefore,

$$S_i''(x) = \frac{(x - x_i)M_{i+1} - (x - x_{i+1})M_i}{x_{i+1} - x_i}.$$

(This is just a linear cross mean.) Integrating this expression twice gives

$$S_i(x) = \frac{M_{i+1}(x - x_i)^3 - M_i(x - x_{i+1})^3}{6h_i} + \alpha_i x + \beta_i, \tag{2.5.7}$$

where $h_i = x_{i+1} - x_i$. Now we fit this expression to the data points (x_i, y_i) and (x_{i+1}, y_{i+1}), i.e.

$$\begin{aligned} M_i(x_i - x_{i+1})^2 + 6\alpha_i x_i + 6\beta_i &= 6y_i, \\ M_{i+1}(x_{i+1} - x_i)^2 + 6\alpha_i x_{i+1} + 6\beta_i &= 6y_{i+1}, \end{aligned}$$

from which we eliminate β_i to produce

$$\alpha_i = \frac{(y_{i+1} - y_i)}{(x_{i+1} - x_i)} - \tfrac{1}{6}(x_{i+1} - x_i)(M_{i+1} - M_i) \tag{2.5.8}$$

and

$$\beta_i = y_i - \alpha_i x_i - \tfrac{1}{6}M_i(x_{i+1} - x_i)^2. \tag{2.5.9}$$

However, $S_i'(x)$ is required to be continuous across x_i, therefore

$$S_{i-1}'(x_i) = S_i'(x_i),$$

i.e.

$$M_i(x_i - x_{i-1}) + 2\alpha_{i-1} = M_i(x_i - x_{i+1}) + 2\alpha_i$$

or

$$\alpha_i - \alpha_{i-1} = \tfrac{1}{2}M_i(x_{i+1} - x_{i-1}). \tag{2.5.10}$$

Finally, we use equation (2.5.8) to eliminate α_i and α_{i-1} from (2.5.10) to give

$$h_{i-1}M_{i-1} + 2(h_{i-1} + h_i)M_i + h_iM_{i+1} = \frac{6(y_{i+1} - y_i)}{h_i} - \frac{6(y_i - y_{i-1})}{h_{i-1}}. \tag{2.5.11}$$

This second-order recurrence relation holds at all internal points, therefore, we have a system of $n-1$ linear equations in the $n+1$ unknowns $M_0, M_1, \ldots, M_n$. In particular this system of equations has a special form since in a matrix notation we obtain entries only on the main diagonal and the diagonals directly above and below this diagonal. These equations are therefore called tridiagonal and special methods have been developed to solve them which take advantage of their structure. (For an introduction to second-order recurrence relations and the solution of tridiagonal systems of linear equations see Harding and Quinney (1986).) The general solution of this recurrence relation will involve two arbitrary values which can be determined if two additional conditions are supplied. There are two particularly popular ways they can be specified: (i) free splines and (ii) clamped splines.

(i) Free splines (natural splines). The simplest way of imposing additional conditions is to set both M_0 and M_n equal to zero. (This is equivalent to assuming that the interpolating function is approximately linear at x_0 and x_n since M_i is the second derivative at x_i.)

Example 2.5.2

Use the program **INTERP** to determine the natural spline through the data points given in table 2.1.1. As there are five points there will be five second derivative values to be determined. Let these values be M_0, M_1, M_2, M_3 and M_4, but because the spline is to be natural we have $M_0 = M_4 = 0$. Next we write down the equation (2.5.11) at the internal points, i.e.

$$\begin{aligned}
0.2M_0+0.8M_1+0.2M_2 &= \frac{6}{0.2}(0.58813-0.39646) - \frac{6}{0.2}(0.39646-0.09956)\\
0.2M_1+0.8M_2+0.2M_3 &= \frac{6}{0.2}(0.77210-0.58813) - \frac{6}{0.2}(0.58813-0.39646)\\
0.2M_2+0.8M_3+0.2M_4 &= \frac{6}{0.2}(0.89608-0.77210) - \frac{6}{0.2}(0.77210-0.58813)
\end{aligned}$$

or

$$\begin{pmatrix} 0.8 & 0.2 & 0 \\ 0.2 & 0.8 & 0.2 \\ 0 & 0.2 & 0.8 \end{pmatrix} \begin{pmatrix} M_1 \\ M_2 \\ M_3 \end{pmatrix} = \begin{pmatrix} -3.156899 \\ -0.231000 \\ -1.799699 \end{pmatrix}.$$

These equations have the solution $M_1 = -4.306179$, $M_2 = 1.440214$ and $M_3 = -2.609679$. These values provide all that is necessary to construct the required natural spline. The resulting function is shown in figure 2.5.6. If we wish to evaluate the spline at a particular point it is first necessary to decide in which sub-interval the required point lies. For example, to evaluate this spline at $x = 0.51$ we need to evaluate the component of the spline which is defined on the interval $(0.5, 0.7)$ or (x_2, x_3). The required function is therefore

$$S_2(x) = \frac{M_3(x - x_2)^3 - M_2(x - x_3)^3}{6(x_3 - x_2)} + \alpha_2 x + \beta_2,$$

where

$$\alpha_2 = \frac{(0.7721 - 0.58813)}{0.2} - \frac{1}{6} \times 0.2(-2.660967 - 1.440214)$$

and

$$\beta_2 = 0.58813 - 0.5\alpha_2 - \frac{1}{6} \times 1.440214(0.2)^2$$

from (2.5.8) and (2.5.9). Evaluating these expressions gives the component of the spline in the interval $(0.5, 0.7)$ and hence $S_2(0.51) = 0.59731$ to five decimal places.

Exercise 2.5.2 (Analytical) Determine an explicit expression for the natural cubic spline through the data points of table 2.1.1. By direct evaluation show that the resulting spline is continuous and has continuous first and second derivatives at the internal points $x_1,\ldots,x_{n-1}$.

Exercise 2.5.3 (Computational) Use the program **INTERP** with the option **Method: Cubic spline - free** to confirm the results of example 2.5.2.

(ii) Clamped splines. If it is possible to supply values for the slope of the interpolating function at x_0 and x_n then we can use equation (2.5.7) to provide the additional equations required. Let us suppose that we are

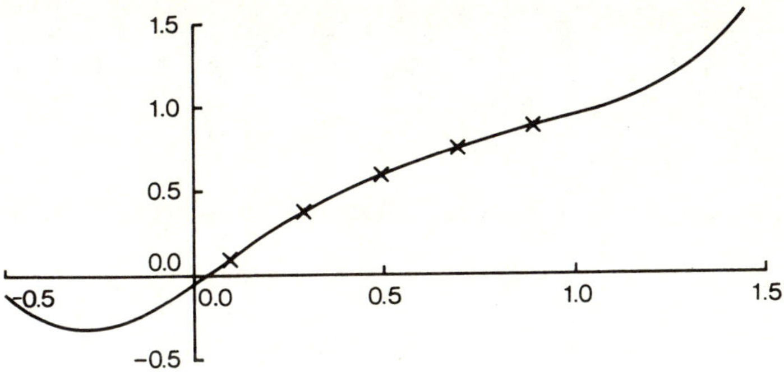

Figure 2.5.6 The free cubic spline for the points in table 2.1.1.

given that $S_0'(x_0) = d_0$ and $S_n'(x_n) = d_n$ then differentiating (2.5.7) and substituting we have

$$d_0 = \frac{(y_1 - y_0)}{h_0} - \frac{h_0}{6}(M_1 + 2M_0)$$

or

$$2h_0M_0 + h_0M_1 = \frac{6(y_1 - y_0)}{h_0} - 6d_0. \tag{2.5.12}$$

and

$$d_n = \frac{(y_n - y_{n-1})}{h_{n-1}} + \frac{h_{n-1}}{6}(M_{n-1} + 2M_n)$$

or

$$h_{n-1}M_{n-1} + 2h_{n-1}Mn = 6d_n - 6\frac{(y_n - y_{n-1})}{h_{n-1}}. \tag{2.5.13}$$

Next we combine equations (2.5.12) and (2.5.13) with the linear equations given by (2.5.11) to obtain $n+1$ linear equations in the unknown values M_i, $i = 0, ..., n$. Once these values are known the required cubic spline is uniquely determined. The matrix system which is formed in this way is both tridiagonal and diagonally dominant and can be solved in a relatively straightforward way by Gaussian elimination without partial pivoting. (See Harding and Quinney (1986).)

Example 2.5.3 (Computational)
Use the program **INTERP** to determine the clamped cubic spline which passes through the data points in table 2.1.1 and has slope 1.4845 at

$x = 0.1$ and 0.6199 at $x = 0.9$. (These values are default settings.) This clamped cubic spline is shown in figure 2.5.7 and can be compared with the free spline in figure 2.5.6.

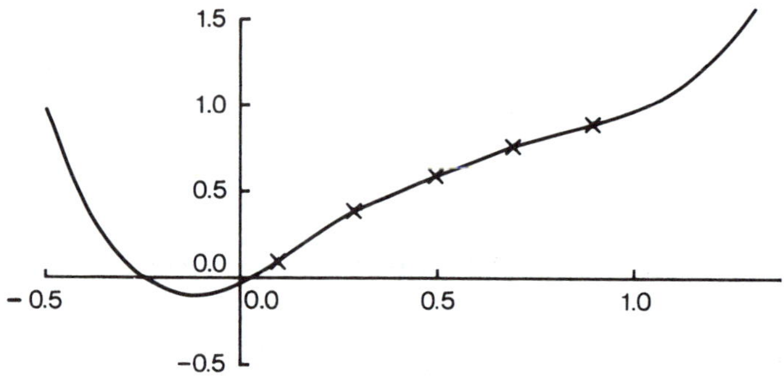

Figure 2.5.7 The clamped cubic spline in example 2.5.3.

Exercise 2.5.4 (Computational) Use the program **INTERP** to determine the clamped cubic spline which passes through the points given in table 2.1.1 and has zero slope at both ends. Compare the result with the free spline and the interpolating polynomial through the same points.

We can now compare the quadratic spline which passes through the points in table 2.1.1, on page 9, which is shown in figure 2.5.4, the clamped cubic spline shown in figure 2.5.7 and the corresponding free cubic spline which is shown in figure 2.5.6. Clearly, the cubic spline is less prone to oscillations and we can show that this is generally the case.

Theorem 2.1. Let $C_2[x_0, x_n]$ be the set of all twice differentiable functions which pass through the tabulation points, (x_i, y_i), $i = 0, \ldots, n$. The natural cubic spline, $S(x)$, which is a member of $C_2[x_0, x_n]$, minimizes the average curvature:

$$\sqrt{\left(\frac{1}{x_n - x_0}\int_{x_0}^{x_n} (f''(x))^2\, dx\right)}.$$

Proof. Let $S(x)$ be the required natural spline given by the solution of (2.5.11) subject to $M_0 = M_n = 0$ and let $f(x)$ be any other twice differentiable function over $[x_0, x_n]$. Now consider

$$\begin{aligned}
\int_{x_0}^{x_n} (f''(x))^2\,dx &= \int_{x_0}^{x_n} [S''(x) + (f''(x) - S''(x))]^2\,dx \\
&= \int_{x_0}^{x_n} S''(x)^2\,dx + 2\int_{x_0}^{x_n} S''(x)[f''(x) - S''(x)]^2\,dx \\
&\quad + \int_{x_0}^{x_n} [f''(x) - S''(x)]^2\,dx
\end{aligned}$$

and then look at the second of these terms, i.e.

$$I = \int_{x_0}^{x_n} S''[f'' - S'']\,dx.$$

Integrating by parts gives

$$\begin{aligned}
I = S''(x_n)[f'(x_n) - S'(x_n)] &- S''(x_0)[f'(x_0) - S'(x_0)] \\
&- \int_{x_0}^{x_n} [f'(x) - S'(x)]S'''(x)\,dx.
\end{aligned}$$

However, for a natural spline $S''(x_0) = M_0 = 0$ and $S''(x_n) = M_n = 0$ and therefore the first two terms vanish. Furthermore, in each subinterval (x_i, x_{i+1}) the cubic spline has constant third derivative which we shall call K_i and so

$$I = -\sum_{i=0}^{i=n-1} \int_{x_i}^{x_{i+1}} [f'(x) - S'(x)]K_i\,dx = -\sum_{i=0}^{i=n-1} K_i[f(x_i) - S(x_i)] = 0.$$

From this we conclude that

$$\int_{x_0}^{x_n} [f''(x)]^2\,dx \geq \int_{x_0}^{x_n} [S''(x)]^2\,dx$$

as required.

In order to demonstrate the importance of this result consider the following example.

Exercise 2.5.4 (Computational) Determine a polynomial approximation for the data in table 2.5.1. (These data points are stored in the file **SPLINE**). Now construct the natural cubic spline which passes through these points and compare the results obtained.

Table 2.5.1 Data for exercise 2.5.4.

x	−1	−0.960	−0.860	−0.690	0.220	0.500	0.930
y	−1	−0.151	0.594	0.986	0.895	0.500	−0.306

Exercise 2.5.5 (Computational) Use the program **INTERP** to determine a quadratic spline through the data in the data file **SPLINE** which has zero slope at its left-hand end, i.e. $d_0 = 0$. Vary d_0 and investigate the behaviour of the resulting spline. Now fit a natural cubic spline through these data points and compare the results. Which curve is the smoothest? Investigate the behaviour of clamped splines for a variety of endpoint conditions. Is it possible to improve upon the natural cubic spline? To see that using a higher order interpolating function does not improve the behaviour, try fitting a high order polynomial using the option **Construct interpolating polyn.** (Hint: the quadratic spline is determined by a first order recurrence relation for which an initial condition is supplied. Is it stable? Why is the two-point boundary value problem formulation involved with cubic splines a more stable process?)

> 2.6 Summary and conclusion

We have seen in this chapter how it is possible to produce a continuous function which passes through a given set of tabular values. The solution to this problem is not unique. We can aim to fit a function which is composed of a linear combination of simple powers of x, i.e. a polynomial, but this may lead to an interpolating polynomial which is highly oscilllatory. Alternatively, we can aim for a more local approximation and fit a spline which will give a resulting function which is less likely to suffer from dramatic oscillations between tabular values. However, whichever method is used we cannot say anything definite about non-tabular values. To illustrate this point see figure 2.6.1. The interpolating function passes through all the tabular values but is likely to be of little use in interpolation.

An underlying assumption with any interpolation problem is that the data points have been determined from some unknown but *continuous* function. If the underlying function is not continuous then no amount of continuous functions can produce satisfactory interpolation near any

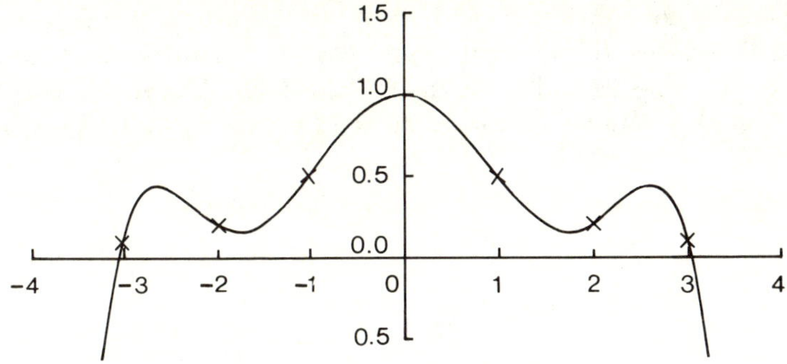

Figure 2.6.1 An example of poor interpolation.

singularity. Chapter 3 will consider this further, but for the moment we conclude this chapter with the following general remarks.

1. Interpolation is usually more reliable than extrapolation; the latter should be avoided if at all possible.

2. Except in all but the very simplest cases it is unwise to determine an interpolating polynomial by considering expressions of the form $a_0+a_1x+a_2x^2+\ldots+a_nx^n$. This problem is possibly ill-conditioned and is likely to become more so as the number of tabular points increases. However, the construction of a polynomial interpolant can be carried out in a stable way using either Lagrangian polynomials or divided differences. If the value of the interpolating polynomial is only required at one or two points then it is better to use Neville's algorithm.

3. Interpolation is a process whereby given a table of values $(x_i, y(x_i))$ and a non-tabular value x^* we attempt to find $y(x^*)$. *Inverse interpolation* is where we are given a value y and are asked to determine a value x which would produce y.

4. The curve which has minimum curvature through a set of points and yet has continuous first and second derivatives is a natural cubic spline. Such a function is made up from a set of cubic functions, one in each tabular interval, which are continuous and have continuous first and second derivatives at each internal tabular point.

We have not considered the idea of rational interpolation, however, we shall consider the rational approximation functions in Chapter 3. Given a set of points and a rational function which passes through them it is clear that rational interpolation can be carried out at non-tabular points.

> Chapter 3

> The Approximation of Functions

'I've got fourteen pots of honey left, or is it fifteen, as the case may be', said Pooh humbly.

'Well, let's call it sixteen' said Rabbit.

With apologies to A A Milne.

In an age of the ever increasing use of pocket calculators and micro-computers, the need to evaluate standard mathematical functions is also increasing. Clearly, electronic devices cannot store all possible values of such functions; the question therefore is how do such machines evaluate expressions such as sin(1.4) or exp(−3.456)? Inherent in such a problem is not only the need to carry out such an evaluation but also the need to know how accurate the presented result is going to be. In this chapter we will examine this question and try to provide some sensible answers.

> 3.1 Taylor series approximation

It is well known that provided a function is sufficiently smooth near a particular point $x = a$ then it is possible to write

$$f(x) = f(a) + (x-a)f'(a) + \frac{(x-a)^2}{2!}f''(a) + \ldots$$

where the infinite series converges for all x which satisfy $|x-a| \leq R$ and R is called the radius of covergence (see Appendix A). This is called a Taylor's series expansion of the function f about the point $x = a$. An alternative form is

$$f(x) = f(a) \quad + \quad (x-a)f'(a) + \frac{(x-a)^2}{2!}f''(a) + \ldots$$

$$\ldots + \frac{(x-a)^n}{n!} f^{(n)}(a) + R_n, \qquad (3.1.1)$$

where the remainder term R_n is given by

$$R_n = \frac{(x-a)^{(n+1)}}{(n+1)!} f^{(n+1)}(\theta), \qquad (3.1.2)$$

and θ lies between x and a. Therefore, if we knew the value of $f(a)$, $f'(a)$, $f''(a)$, ..., $f^{(n)}(a)$, we could use equation (3.1.1) to estimate $f(x)$. Furthermore, the maximum possible error of such an approximation is given by $\max|R_n|$ for all values of θ between x and a.

Example 3.1.1
We can use a Taylor series with $a = 0$, which is sometimes called a Maclaurin series, to estimate exp(0.1). We can then consider questions such as the following. If five terms of this series are used what is the maximum possible error? How many terms are required to evaluate exp(0.5) correct to six decimal places? The Taylor series expansion of $\exp(x)$ about $x = 0$ is given by

$$\exp(x) = 1 + x + \frac{x^2}{2!} + \frac{x^3}{3!} + \frac{x^4}{4!} + \ldots + \frac{x^n}{n!} + \ldots.$$

Taking the first five terms gives

$$\exp(0.1) \approx 1 + 0.1 + \frac{0.1^2}{2} + \frac{0.1^3}{6} + \frac{0.1^4}{24},$$

therefore, $\exp(0.1) \approx 1.105170831$ which can be compared with the correct value of $\exp(0.1) = 1.10517092\ldots$. The remainder term R_4 associated with the above Taylor series approximation may be used to estimate the error involved with this approximate value. From equation (3.1.2) with $x = 0$ and $a = 0.1$ we have that

$$|R_4| \leq \max_{0 \leq x \leq 0.1} \left| \frac{0.1^5 \exp(\theta)}{120} \right| \leq 0.1^5 \frac{\exp(0.1)}{120} \leq 9.2 \times 10^{-8},$$

compared with the actual error of 9.0×10^{-8}. Suppose we now wanted to find exp(0.5) using the same quartic polynomial expression then

$$\exp(0.5) \approx 1 + 0.5 + \frac{0.5^2}{2} + \frac{0.5^3}{6} + \frac{0.5^4}{24} = 1.6484374$$

but this compares less favourably with $\exp(0.5) = 1.6487213\ldots$. For this value of x, $\max|R_4| = 4.29\times10^{-4}$. Clearly, the range of values for which we can use this polynomial approximation is limited, but one way around this problem would appear to be to include further terms. Alternatively, we can use the remainder term, R_n, to see just how many terms are required. For example, in order to ensure that the approximation has a maximum error, for all $|x| \le \frac{1}{2}$, which is less than 1 part in 10^6 we need to select an n such that

$$|R_n| \le 0.5 \times 10^{-7}, \text{ for all } |x| \le \tfrac{1}{2}.$$

Therefore, from (3.1.2) we need to find an integer n such that

$$\max_{|x|\le\frac{1}{2}} \left| \frac{0.5^{n+1}\exp(x)}{(n+1)!} \right| \le 0.5 \times 10^{-7}$$

which can be achieved if $n \ge 9$. This result implies that if we require an accuracy of six decimal places each time we evaluate the exponential function for $|x| < \frac{1}{2}$ then a polynomial of degree at least 9 is required. Clearly, for larger values of x more and more terms would be needed.

Example 3.1.2 (Computational)
We shall demonstrate a Taylor series approximation for the function $\exp(x)$, $|x| \le 1$, using the program **APPROX**. (The program has a default function setting for this problem.) Load the program **APPROX** and when the prompt **Function** is shown accept the default value **as set** which will produce a new prompt **Method**. Cycle through the options available by repeatedly pressing the SPACE BAR and when the option shows **Taylor Series** press RETURN. Again cycle through the options available, to familiarize yourself with the program, and when it shows **Option: calculate coeffs** press RETURN again. Now set the value of **degree** to 3, which will ask for a cubic approximation, and finally put $a = 0$. After a short pause the prompt will change to **Plot: function**; press RETURN to accept this. The prompt should now be **Graph data: superimpose**; however, clear the screen by changing this option to **Graph data: clear – same axes** and then press RETURN. The function will now be plotted in the graphics window and then the program will return to the option **Plot: function**. Now change the **Plot** option to **approximation** by pressing the SPACE BAR and press RETURN to show the required cubic Taylor series approximation for $\exp(x)$ on this interval. It is also possible to plot the error between the function and approximation by changing the plotting option, **Plot**, to show **Plot: error**. The

result is shown in figure 3.1.1. The associated error function is plotted on the same x axis but the corresponding y axis is scaled by a suitable factor which is displayed on the screen. In order to find this approximation it is necessary to determine suitable values for the coefficients in equation (3.1.1) and with the option **calculate coeffs** this is done by using finite difference approximations. (See Appendix B.) Exercise 3.1.1 demonstrates how it is possible for the user to supply these coefficients manually.

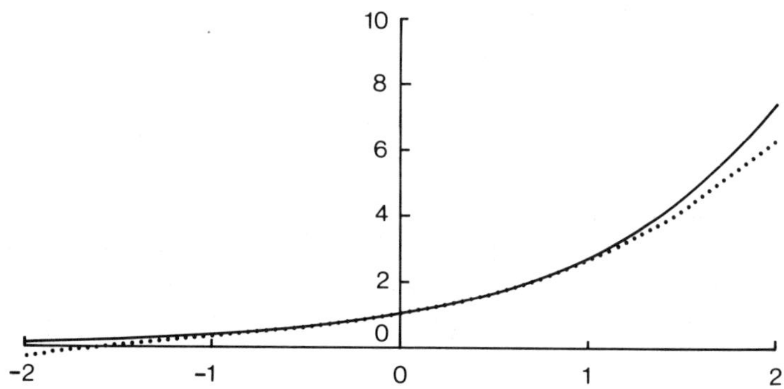

Figure 3.1.1 Taylor series expansion for $\exp(x)$ about $x = 0$.

Exercise 3.1.1 (Computational) Load the program **APPROX** and the function $\exp(x)$, $|x| \leq 1$. Select the **Method: Taylor Series** and the **Option: calculate coeffs** and then plot the function, its cubic Taylor series approximation and the associated error as described in example 3.1.2. Exit from this part of the program by taking the option **Plot: quit** and follow the instructions given. This will return the program to the prompt **Option: calculate coeffs**. Again cycle through the options available at this point which include the possibility of storing the current coefficients on a suitable disc file, reading another set of coefficients which you have previously stored, or **set coefficients manually**. The degree of the current approximation should be 3 and the point about which the expansion is determined should be $a = 0$; if not reset these values. The correct values for the first four coefficients in the Taylor series of $\exp(x)$ about $x = 0$ are 1, 1, $\frac{1}{2}$ and $\frac{1}{6}$. Accept the option to **set coefficients manually** and then input these values; notice that the current values, which are determined by finite differences, are shown on the screen and are very close to the correct values. Compare the approximations which are determined.

Exercise 3.1.2 (Computational) Use the program **APPROX** to determine a cubic Taylor series expansion about $x = 0$ for $\exp(x)$, $|x| \leq 2$. Compare the results with a corresponding expansion about the points $x = 1$ and $x = -1$.

Exercise 3.1.3 (Analytical) How many terms are required to evaluate the function $\log(1 + x)$, $|x| < \frac{1}{2}$, correct to eight decimal places?

As the evaluation of polynomials plays a fundamental role in this type of approximation it is appropriate to consider the most efficient way of carrying out this task. We shall illustrate a technique called *nested multiplication*. This method permits the evaluation of a polynomial expression without the storage of intermediate results and therefore it can be carried out on a hand calculator which does not have a memory.

Exercise 3.1.4 (Analytical) Polynomial evaluation by *nested multiplication*. Let $p_n(x) = a_n x^n + a_{n-1}x^{n-1} + \ldots + a_1 x + a_0$. Determine the number of multiplications and additions required to evaluate this polynomial directly for a fixed value of x. Show that if we define a sequence u_j by

$$u_n = a_n, \; u_{j-1} = xu_j + a_{j-1}, \qquad j = n, n-1, \ldots, 1, \tag{3.1.3}$$

then $p_n(x) = u_0$. This is called nested multiplication. Determine the number of operations required to evaluate the polynomial p_n using this method. Which method is the most efficient? Notice that by using nested multiplication $p_n(x)$ can be calculated without any intermediate memory.

For a fixed value of x the technique of nested multiplication introduced in exercise 3.1.4 is equivalent to a first-order recurrence relation. However, the recurrence relation (3.1.3) is stable only if $|x| \leq 1$ (see Harding and Quinney (1986)). To demonstrate the importance of this comment let us suppose that we wish to approximate $\sin(x)$, for a value of x in the interval $[10, 11]$, by using a Taylor series expansion of degree 3 about the point $x = 10$. This gives

$$p_3(x) = \sin(10) + (x-10)\cos(10) - \tfrac{1}{2}(x-10)^2\sin(10) - \tfrac{1}{6}(x-10)^3\cos(10). \tag{3.1.4}$$

Expanding out this expression gives the cubic

$$p_3(x) = a_0 + a_1 x + a_2 x^2 + a_3 x^3 \text{ for } x \in [10, 11], \tag{3.1.5}$$

where $a_0 = -104.798$, $a_1 = 35.6743$, $a_2 = -3.92335$ and $a_3 = 0.139845$, correct to six significant figures. Now suppose we wish to find $p_3(10.5)$. As a_3 is accurate to six decimal places, when it is be multiplied by a factor $(10.5)^3$ the resulting value of $p_3(10.5)$ may only have two or possibly three decimal places correct. Clearly, the evaluation of the polynomial (3.1.5) may be prone to induced errors and should be avoided. Instead we leave it in the form (3.1.4), i.e.

$$\begin{aligned} p_3(x) = -0.544021 - 0.839072(x-10) &+ 0.272011(x-10)^2 \\ &+ 0.139845(x-10)^3 \end{aligned}$$

which ensures that all the terms raised to powers are less than 1; hence we should have a stable recurrence relation and reasonable conditioning. This introduces a very important point, for clearly this problem will always arise if we attempt to approximate any function outside the interval either $[0,1]$ or $[-1,1]$. However, we can always scale any interval $[a,b]$ to either of these standard intervals by a suitable linear scaling procedure.

Exercise 3.1.5 (Calculator) By examining the possible errors in the coefficients of the polynomial (3.1.5) determine *upper and lower error functions* for this expression.

Exercise 3.1.6 (Analytical) Scale the interval $[2,10]$ to the standard intervals $[0,1]$ and $[-1,1]$.

> 3.2 Polynomial approximation

We have seen how it is possible to obtain a polynomial approximation for a given function by truncating a corresponding Taylor series expansion. Now we will consider an alternative approach using interpolating polynomials. Let us suppose that we are given a function $f(x)$, x in (a,b), and attempt to find a polynomial function of degree n, $p_n(x)$, which is close to f in some sense.

Example 3.2.1 (Computational/Calculator)
Let us determine a polynomial approximation of degree 3 for the function $f(x) = \exp(x)$, for values of x in $[-1,1]$. We begin by considering the Taylor series expansion of $\exp(x)$ in this interval. This is given by

$$\exp(x) = 1 + x + \tfrac{1}{2}x^2 + \tfrac{1}{6}x^3 + R_3(x),$$

where the remainder term satisfies $|R_3(x)| \leq \max_{|x|\leq 1} |f^{iv}|/4! \leq 0.113$. This approximation is shown in figure 3.2.1 and the error function, $R_3(x)$, in figure 3.2.2. As an alternative method we can make the function and approximating polynomial agree at a given set of points, i.e. we select the coefficients a nth degree polynomial so that $p_n(x_i) = f(x_i)$, $i = 0, ..., n+1$. Thus the problem becomes one of constructing an interpolating polynomial through $n+1$ points. Firstly, we need to decide where we are going to put the tabulation points x_i, $i = 0, 1, 2, 3$, at which the polynomial and function agree. For the moment, let us put them at equally spaced points in $[-1, 1]$, i.e. $-1, -\frac{1}{3}, \frac{1}{3}$ and 1, and assume that the required cubic polynomial is

$$p_3(x) = a_0 + a_1 x + a_2 x^2 + a_3 x^3.$$

Then at $x = -1$ we require that

$$p_3(-1) = a_0 - a_1 + a_2 - a_3 = \exp(-1).$$

We obtain a similar equation at the remaining three points which gives a system of four linear equations in the unknown coefficients a_0, a_1, a_2 and a_3 (see section 2.1). These equations can be solved using Gaussian elimination to produce the cubic approximation

$$p_3(x) = 0.995196 + 0.999049x + 0.547885x^2 + 0.176152x^3. \qquad (3.2.1)$$

This function is also plotted in figure 3.2.1 and its error in figure 3.2.2. Notice that the maximum error associated with the polynomial (3.2.1) is much less than the maximum error associated with the corresponding Taylor expansion of the same order.

We have already seen how it is possible to estimate the error associated with a Taylor series approximation of a given function. Now we will construct an error function for polynomial interpolation. Firstly, we recall that $f(x)$ and the interpolant $p_n(x)$ agree at the $n+1$ points x_i, $i = 0, 1, \ldots, n$, therefore,

$$e_n(x) = f(x) - p_n(x) = K(x)(x - x_0)(x - x_1)\ldots(x - x_n),$$

where $K(x)$ is to be determined. For any given value of x let us define the function ϕ by

$$\phi(t) = f(t) - p_n(t) - K(x)w(t),$$

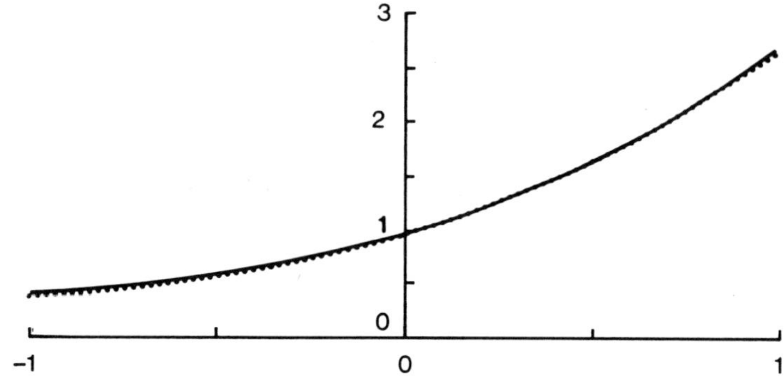

Figure 3.2.1 Taylor series approximation and $p_3(x)$ for $\exp(x)$.

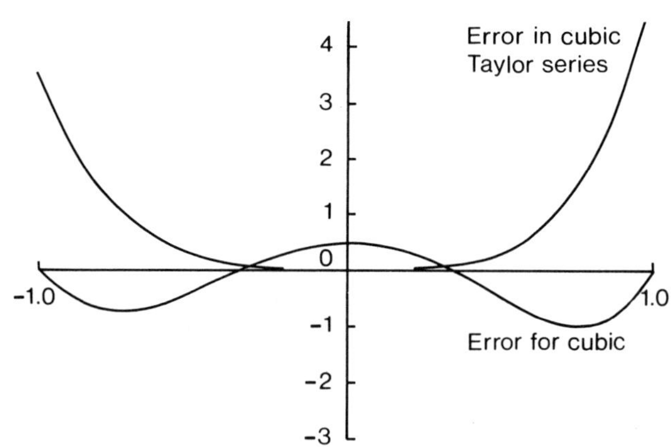

Figure 3.2.2 Error functions associated with figure 3.2.1.

where $w(t) = (t-x_0)(t-x_1)\ldots(t-x_n)$, and then notice that ϕ vanishes at the points $n+2$ points x, x_0, ...,x_n. By Rolle's theorem ϕ' must vanish between any two points at which ϕ vanishes and so ϕ' has $n+1$ zeros. Similarly, ϕ'' has n zeros and so on until we find that the $(n+1)$th derivative of ϕ, $\phi^{(n+1)}$, has exactly one zero; let this zero be θ, then

$$0 = \phi^{(n+1)}(\theta) = f^{(n+1)}(\theta) - p_n^{(n+1)}(\theta) - Kw^{(n+1)}(\theta).$$

However, $p_n^{(n+1)}$ is identically zero and $w^{(n+1)} = (n+1)!$, therefore,

$$K = \frac{f^{(n+1)}(\theta)}{(n+1)!},$$

or

$$f(x) - p_n(x) = (x-x_0)(x-x_1)\ldots(x-x_n)\frac{f^{(n+1)}(\theta)}{(n+1)!}. \qquad (3.2.2)$$

(Recall that θ will depend upon the choice of x.) Notice the similarity between the form of this error function and that associated with the Taylor series expansion given by equation (3.1.2).

Example 3.2.2 (Calculator/Computer)
We shall determine an upper bound on the error associated with the cubic approximation for $\exp(x)$ which was derived in example 3.2.1, i.e.

$$p_3(x) = 0.995196 + 0.999049x + 0.547885x^2 + 0.176152x^3.$$

From equation (3.2.2) with $n = 3$ we have that

$$|f(x) - p_3(x)| \le |(x+1)(x+\tfrac{1}{3})(x-\tfrac{1}{3})(x-1)|M/4!,$$

where $M = \sup|f^{iv}|$, $x \in [-1,1]$, and $f(x) = \exp(x)$. (We define the supremum as follows: if $M \ge |f^{iv}|$ and $\forall\ L \ge |f^{iv}|$, $M \le L$ then M is the supremum.) We now examine the first four terms and note that for values of x between -1 and 1 this takes a maximum value of 0.19753 at $x = \pm\frac{1}{3}\sqrt{5}$. (See figure 3.2.3.) Therefore, since $|f^{iv}(x)| = |e^x| \le e$, for $x \in [-1,1]$, the error function $e_3(x)$ satisfies $|e_3(x)| \le 0.0224$, which is the required error bound. In fact this is a severe overestimate since the maximum error is actually 0.00481 but even a crude bound is better than none. (Recall that the corresponding bound on the cubic Taylor series approximation was $|R_3(x)| \le 0.113$.)

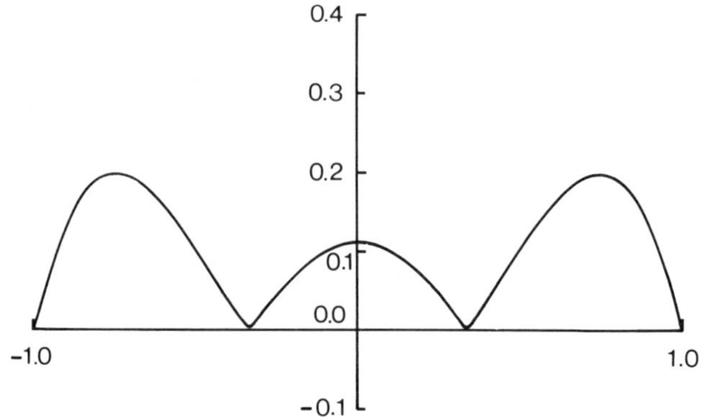

Figure 3.2.3 $y = |(x+1)(x+\frac{1}{3})(x-\frac{1}{3})(x-1)|$, $|x| \leq 1$.

Exercise 3.2.1 (Analytical) By determining upper and lower bounds on the fourth derivative of $\exp(x)$ construct upper *and* lower bounds for the error function $e_3(x)$ given in example 3.2.1.

Exercise 3.2.2 (Computational) Use the program **APPROX** to determine a polynomial approximation of order 2 for $\exp(x)$, $|x| \leq 1$, by interpolation at the points $x = -1, -\frac{1}{3}$ and 1. Load the program **APPROX**, accept the option **Function as set** and then select **Method: Interpolation**. Now select **Option: edit/inspect points** and enter the three points on the curve at $x = -1, -\frac{1}{3}$ and 1. (When you have selected this option you will be given the opportunity to change the plotting ranges if you wish to do so.) Select the editing option, cycle through the edit options until it shows **Add Point** and then press RETURN. The top window will now contain information regarding how to input a point and, in the graphics window, there should be a flashing cursor at the point $(-1, 0)$. For example, if you press the COPY key the cursor will move to the point $(-1, \exp(-1))$. If you press RETURN this value will be accepted as an interpolation point and the number of points, N, which is shown in the bottom window will be increased by 1. Finally, the edit option will be returned to **Add Point**. Now input two additional points at $x = -\frac{1}{3}$, $y = \exp(-\frac{1}{3})$ and $x = 1$, $y = \exp(1)$. The x values can either be input by pressing 'X' or by using the cursor which is controlled in the usual way. (Each key press will move the cursor by an amount 'dX' and this variable is in turn controlled by the '>' and '<' keys.) When you are satisfied with the points which have been input quit the editor and select **Option: calculate coeffs** then press RETURN. The computer will now determine the polynomial through these points, display the resulting polynomial in the bottom window and then display the prompt **Plot: function**. By changing the prompt you can plot the approximation or the error between the function and the computed polynomial. Use the editor to change the interpolation points and investigate how the approximation and its error change. Now return to the editor and add another two points and then determine quartic polynomial approximations for $\exp(x), x \in [-1, 1]$.

Exercise 3.2.4 (Computational) Use the program **APPROX** with the option **Method: Interpolation** to determine the cubic polynomial approximations for the function $\exp(x)$, x in $[-1, 1]$, by interpolation at -1, -0.5, 0.5 and 1. Investigate how the error between the interpolating polynomial and the function changes as the tabular points change. Investigate how the error changes as the number of tabular points increases. Find the polynomial of smallest degree which approximates this function to an accuracy of at worst 0.001 throughout this interval for interpolation points which are equally spaced. Investigate how the error function behaves for different interpolation points.

We have assumed that it is always possible to construct a suitable polynomial approximation, but is this always the case? Fortunately, provided the function f satisfies certain minimal requirements then there is always such a polynomial. We state, without proof, the following theorem.

Theorem 3.2.1. Weierstrass's approximation theorem. If f is a continuous function on the finite interval $[a, b]$, then for any $\epsilon > 0$ there exists an nth degree polynomial $p_n(x)$, *for some* n, such that

$$|f(x) - p_n(x)| < \epsilon \quad \text{for all } x \in [a, b].$$

In other words, if we are given a continuous function f, defined on the interval $[a, b]$, then there is always a polynomial function which will approximate f to within any prescribed error ϵ. Unfortunately, what Weierstrass's theorem does not tell us is how large n should be, nor how to find the associated polynomial. Furthermore, this theorem assumes that we can carry out all arithmetic exactly, which is not the case either.

> 3.3 Minimax polynomials

In the last section we saw that there are many polynomial approximations of a given degree for any given function. For example,

$$\exp(x) \approx 1 + x + \tfrac{1}{2}x^2 + \tfrac{1}{6}x^3, \tag{3.3.1}$$

and

$$\exp(x) \approx 0.99520 + 0.99905x + 0.54788x^2 + 0.17615x^3, \tag{3.3.2}$$

are both cubic approximations. The first is a truncated Taylor series, the second is an interpolating polynomial which agrees with $\exp(x)$ at the points -1, $-\frac{1}{3}$, $\frac{1}{3}$ and 1. We can compare these functions as shown in figure 3.2.1 but it is more informative to look at their respective error functions shown in figure 3.2.2. If we were to obtain another interpolating cubic approximation by adjusting the tabular points we could plot the associated error function in the same way and compare it. In general we need to find a polynomial, $p_n(x)$, which minimizes the maximum value of the error function, i.e. for any other polynomial of degree n, $q_n(x)$ say, the minimax error ϵ_n satisfies

$$\epsilon_n = \max|f(x) - p_n(x)| \leq \max|f(x) - q_n(x)| \qquad \text{for all } x \in [a, b].$$

The resulting polynomial is called the *minimax polynomial of degree n*. By adjusting the tabular points at which a polynomial $q_n(x)$ agrees with $f(x)$ we can look for the minimax polynomial, but this is likely to be a long process. Instead we can use a very powerful result due to Chebyschev.

Theorem 3.3.1. Chebyshev's equi-oscillation theorem. Let $f(x)$ be a continuous function defined on the interval $[a, b]$ and $p_n(x)$ be a polynomial approximation, of degree n, for f on $[a, b]$ with associated error function $e_n(x) = f(x) - p_n(x)$. If there exists a set of at least $n + 2$ points, $a \leq x_0 < x_1 < \ldots < x_{n+1} \leq b$, where the error function takes maximum values such that

$$e_n(x_i) = (-1)^i \epsilon_n, \quad i = 0, 1, \ldots, n+1,$$

then $p_n(x)$ is the minimax approximation. This approximation is the only one of degree $\leq n$ with this property.

Notice that the equi-oscillation theorem only requires alternating minima and maxima in the error function; they need not be turning points as we shall see in the next example. This theorem is extremely important for it enables us to decide whether or not a particular polynomial is minimax amongst all polynomials of a given degree. For example, the error associated with the third-order approximation (3.3.2), which is shown in figure 3.2.2, has five alternating minima and maxima at the points P, Q, R, S and T, but since they are not all of the same magnitude this polynomial is not minimax.

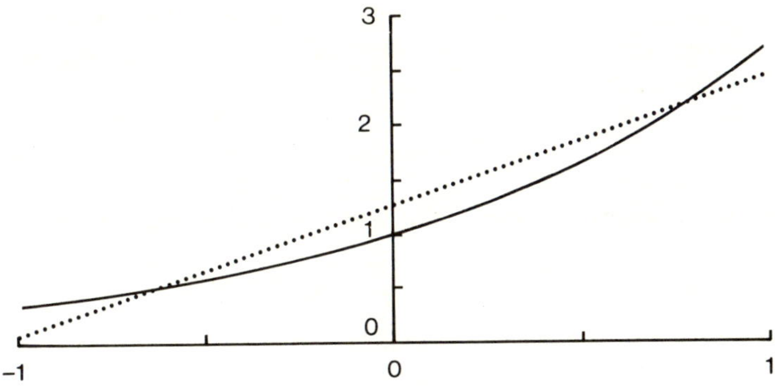

Figure 3.3.1 Linear minimax approximation of $\exp(x)$, $|x| \leq 1$.

Example 3.3.1 (Analytical)
In order to demonstrate the minimax equi-oscillation property we will construct a minimax linear approximation for $\exp(x)$ on the interval $[-1, 1]$. Let us write the linear minimax approximation for $\exp(x)$ as $p_1(x) = ax + b$, then the error function $e_1(x)$ is

$$e_1(x) = \exp(x) - ax - b.$$

As the degree of the polynomial approximation is 1 we need to determine three points x_0, x_1 and x_2 such that $e_1(x)$ takes maximum absolute values which are of the same size but alternating sign, i.e.

$$e_1(x_0) = E, \; e_1(x_1) = -E, \text{ and } e_1(x_2) = E.$$

At least one of x_0, x_1 and x_2 must be an internal point of the interval $[-1, 1]$ at which there will be a turning point, i.e. $de_1/dx = 0$, which gives

$$\exp(x) - a = 0.$$

This equation has only one solution, therefore only one of x_0, x_1 and x_2 is a turning point. Let this be x_1. Accordingly, $e_1(x)$ will have maximum absolute value at both x_0 and x_2 neither of which will be turning points. Therefore,

$$\begin{aligned}
\exp(-1) + a - b &= E \\
\exp(x_1) - ax_1 - b &= -E \\
\exp(1) - a - b &= E
\end{aligned}$$

which gives three equations in the four unknowns; the unknown point x_1, the coefficients a and b, and the error E. However, at x_1 the error function has a turning point and so

$$\frac{de_1}{dx}(x_1) = \exp(x_1) - a = 0,$$

which gives a fourth equation. The solution of these equations is:

$$\begin{aligned}
a &= \frac{(\exp(1) - \exp(-1))}{2} = 1.17520119, \\
x_1 &= \log_e(a) = 0.161439361, \\
b &= \frac{(\exp(x_1) + \exp(1) - ax_1 - a)}{2} = 1.26427905 \\
E &= \exp(1) - a - b = 0.278801586.
\end{aligned}$$

The resulting polynomial approximation is shown in figure 3.3.1 together with its error function in figure 3.3.2. Recall that we are seeking a polynomial of degree 1 and notice that the error function in figure 3.3.2 has three alternating minima and maxima at P, Q and R, all of which are equal to E in magnitude. Therefore, this function is the linear minimax approximation for $\exp(x)$, for x in $[-1, 1]$ and E is the minimax error ϵ_1. In order to investigate this idea further consider an alternative linear approximation for $\exp(x)$ given by $q_1(x) = 1 + x$, the error function for which is also shown in figure 3.3.2. It is clear that the maximum error of q_1 is greater than that of p_1. The Chebyshev equi-oscillation theorem states that for any other linear approximation the associated error has a maximum value greater than the error associated with the minimax linear approximation $p_1(x)$. For linear minimax approximations it is possible to determine the correct expression algebraically, but in general this is not the case.

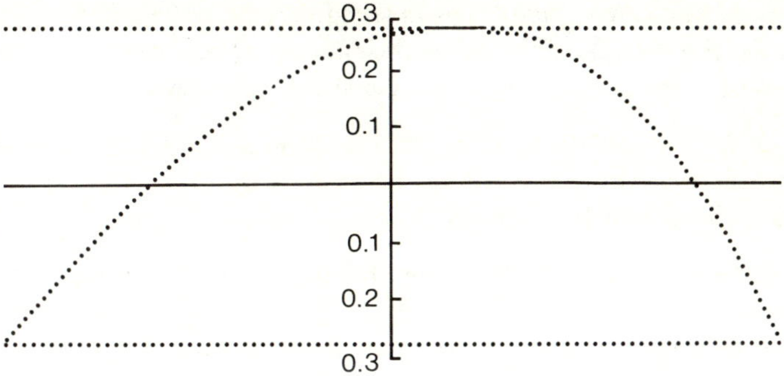

Figure 3.3.2 Linear minimax error function for $\exp(x)$, $|x| \leq 1$.

Example 3.3.2 (Analytical/Computational)
We shall attempt to construct a cubic minimax approximation for $\exp(x)$ for x in $[-1, 1]$. Firstly, let us examine the problems of determining the minimax approximation analytically. We need to find a set of five points, ($n = 3$ since we seek a cubic approximation), x_0, x_1, x_2, x_3 and x_4, where the error function

$$e_3(x) = \exp(x) - a_3x^3 - a_2x^2 - a_1x - a_0$$

has alternating minima and maxima *all of which are of equal magnitude.* Differentiating this function shows that de_3/dx has no more than three zeros in $[-1, 1]$ and so $x_0 = -1$ and $x_4 = 1$ at which the error function will have maximum absolute value. (To see this, sketch the curve $y = \exp(x)$ and any other cubic.) Therefore, we have to solve the equations

$$\begin{aligned}
\exp(-1) + a_3 - a_2 + a_1 - a_0 &= E \\
\exp(x_1) - a_3x_1^3 - a_2x_1^2 - a_1x_1 - a_0 &= -E \\
\exp(x_2) - a_3x_2^3 - a_2x_2^2 - a_1x_2 - a_0 &= E \\
\exp(x_3) - a_3x_3^3 - a_2x_3^2 - a_1x_3 - a_0 &= -E \\
\exp(1) - a_3 - a_2 - a_1 - a_0 &= E
\end{aligned} \qquad (3.3.3)$$

plus the conditions for turning points at x_1, x_2 and x_3, i.e.

$$\begin{aligned}
\exp(x_1) - 3a_3x_1^2 - 2a_2x_1 - a_1 &= 0 \\
\exp(x_2) - 3a_3x_2^2 - 2a_2x_2 - a_1 &= 0 \\
\exp(x_3) - 3a_3x_3^2 - 2a_2x_3 - a_1 &= 0
\end{aligned}$$

which gives a system of eight *non-linear* equations in the unknown values x_1, x_2, x_3, the error E, and the coefficients a_0, a_1, a_2 and a_3. In general it would be extremely difficult to solve these equations.

Exercise 3.3.1 (Analytical) Construct a cubic minimax approximation for the function $\cos(x)$ for values of x in $[-1,1]$. (Hint: use symmetry to reduce the size of the problem.)

Example 3.3.2 demonstrates the difficulty in constructing minimax approximations of order $n > 1$. We will now consider an alternative iterative approach. Consider the cubic approximation for $\exp(x)$ given by (3.2.1), i.e.

$$\exp(x) \approx 0.99520 + 0.99905x + 0.54788x^2 + 0.17615x^3,$$

which is shown in figure 3.2.1. The associated error, as shown in figure 3.2.2, has the correct number of minima and maxima but they are of different magnitude. The internal stationary points of the error function occur at approximately -0.7, 0 and 0.8, therefore let us solve the equations (3.3.3) with $x_1 = -0.7$, $x_2 = 0$ and $x_3 = 0.8$. This produces the *linear* equations

$$\begin{pmatrix} 1 & -1 & 1 & -1 & 1 \\ 1 & -0.7 & 0.7^2 & -0.7^3 & -1 \\ 1 & 0 & 0 & 0 & 1 \\ 1 & 0.8 & 0.8^2 & 0.8^3 & -1 \\ 1 & 1 & 1 & 1 & 1 \end{pmatrix} \begin{pmatrix} a_0 \\ a_1 \\ a_2 \\ a_3 \\ E \end{pmatrix} = \begin{pmatrix} \exp(-1) \\ \exp(-0.7) \\ \exp(0) \\ \exp(0.8) \\ \exp(1) \end{pmatrix}$$

and in turn the solution of these equations gives the cubic

$$q_3(x) = 0.99465 + 0.99643x + 0.54308x^2 + 0.17877x^3. \qquad (3.3.4)$$

The error $e_3(x) = \exp(x) - q_3(x)$ is shown in figure 3.3.3, together with the computed value for E, and has turning points at -0.687312, 0.045814 and 0.727528 in addition to the maxima at -1 and $+1$. (These extremum points are denoted by P, Q, R, S and T). As we seek a minimax approximation of degree 3 the error should have 3+2 alternating extrema of the same magnitude. From figure 3.3.3 we can see that the error function does have five extrema which have alternating sign but they are of different magnitude. If the extrema were of the same magnitude then E would be the minimax error ϵ_3 for $\exp(x)$. We could now substitute these five points into (3.3.3) to determine another approximation and hence iteratively refine our cubic approximation. (Notice that only the x co-ordinate is required to determine the next approximation.) This is the basis of the *Remes algorithm* for determining minimax polynomial approximations which can be described schematically as in figure 3.3.4.

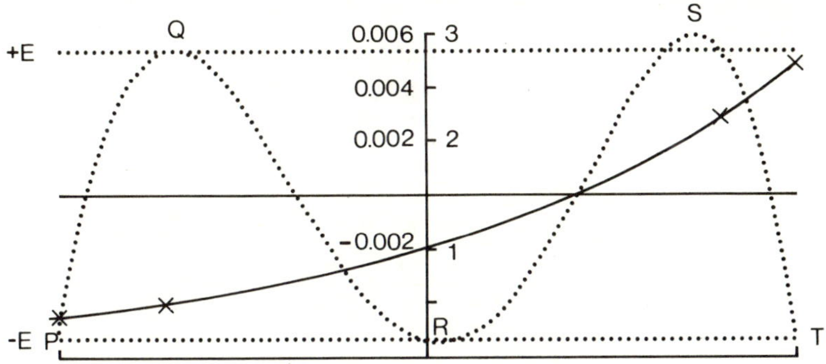

Figure 3.3.3 Error curve associated with the cubic (3.3.4).

Example 3.3.3 (Computational)
We demonstrate the Remes algorithm by considering a cubic minimax approximation for $\exp(x)$, $-1 \leq x \leq 1$. Load the program **APPROX**, for which this is the default function. When the prompt shows **Function: as set** press RETURN and select the option **Method: Minimax/Remes**. As we seek the minimax approximation of order 3, we will need to supply five points at which the required approximation for $\exp(x)$ has alternating minima and maxima. For the moment let us use the values $x = -1, -0.7, 0, 0.8$ and 1 which can be input using the prompt **Option: edit/inspect points**. When you are satisfied that suitable points have been entered quit this option and change the prompt to **Option: calculate coeffs**. (It is also possible to supply a first approximation for the minimax polynomial using the option **Taylor Series**). Now press RETURN and the computer will determine the solution of the equations in step 2 of the Remes flowchart, shown in figure 3.3.4, and display the prompt **Plot: function**. If the function has not been plotted prior to this point then pressing RETURN will display it in the graphics window, otherwise, change the prompt to **Plot: approximation** and compare the computed approximation with the function. It is likely that the function and the computed approximation are very close; the error between them

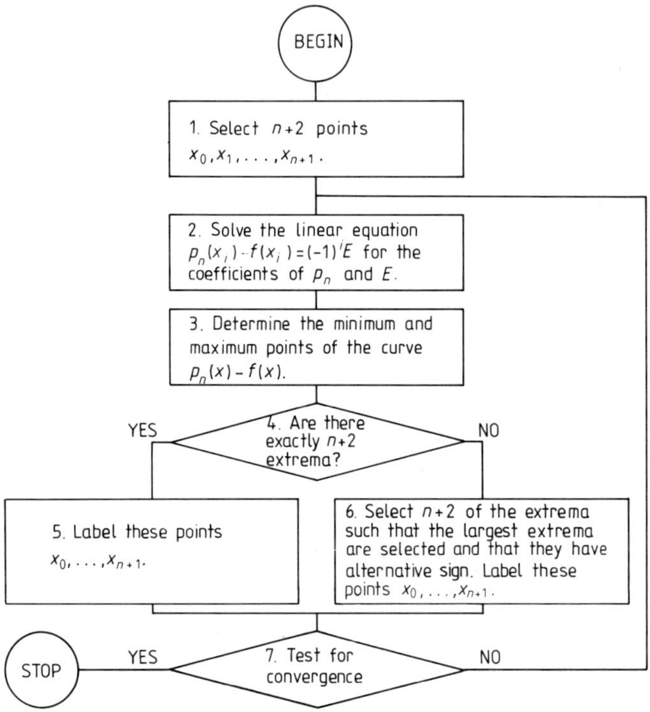

Figure 3.3.4 The Remes algorithm.

can be examined by using the option **Plot: error**. For this example you should obtain figure 3.3.3 together with a dotted line which denotes the value for E in equations (3.3.3). The coefficients of the computed approximation can be examined by using prompt **Option: tabulate**, which in this case gives the cubic (3.3.4). Notice that the error associated with this approximation has maximum values at the points P, Q, R, S and T. We can select these points to be used in the next stage of the Remes algorithm by using the prompt **Option: edit/inspect points** and then return to **Option: calculate coeffs** to determine the next approximation. (Only the x coordinates are required, therefore, it is unnecessary to enter the y coordinates.) This process can then be repeated until convergence is obtained.

Exercise 3.3.2 (Computational) Use the program **APPROX** to carry out the next iteration of the Remes algorithm using the minima/maxima of the error function associated with the cubic approximation (3.3.4).

Exercise 3.3.3 (Computational) Use the program **APPROX** with the option **Remes/minimax** to determine the quartic minimax approximation for $\exp(x), |x| \leq 1$.

Exercise 3.3.4 (Computational) Use the program **APPROX** with the option **Remes/minimax** to determine the quartic minimax approximation for $\log(1+x), |x| \leq \frac{1}{2}$. Try different starting polynomial approximations to examine the speed at which this algorithm converges to the required minimax approximation.

At each stage of the Remes algorithm it is to be expected that the maximum value of the associated error function will be reduced but it will certainly be an upper bound on the minimax error ϵ_n for polynomial approximations of order n. In fact we can also produce a *lower* bound for the minimax error ϵ_n.

Theorem 3.3.2. (de la Vallee Poussin).
Given that q_n is an nth degree polynomial approximation for a function $f(x), x \in [a,b]$, and x_i, $i = 0, \ldots, n+1$, are any $n+2$ points in $[a,b]$ such that $a \leq x_0 < x_1 < \ldots < x_{n+1} \leq b$ where the error function $e_n(x) = f(x) - q_n(x)$ alternates in sign at successive points, then

$$\min_{x_i} |e_n(x_i)| \leq \epsilon_n \leq \max_{x \in [a,b]} |e_n(x)|,$$

where ϵ_n is the minimax error.

Notice that the upper bound in theorem 3.3.2 is over the whole interval, but that the lower bound is only at the points $x_0, x_1, \ldots, x_{n+1}$. For example, the cubic approximation (3.3.4) produces the bounds

$$0.0053526 \leq \epsilon_n \leq 0.0059120.$$

(The lower bound is taken over the points -1, -0.7, 0, 0.8 and 1.0 and occurs at $x = -0.7$. The upper bound is over the whole of the interval and occurs at $x \approx 0.7275$.) If these bounds are sufficiently close to each other then there is little point in carrying out another iteration of the Remes algorithm. What is more, the lower bound tells us that if we wish to have a minimax error less than 0.0053526 then we must construct a polynomial approximation of higher order.

The Remes algorithm is based upon the selection of exactly $n+2$ alternating minima/maxima from a given error function, but it is possible to obtain an error curve for a polynomial approximation of degree n which has more than $n+2$ such points. (Although $p_n(x)$ cannot have more than $n-1$ turning points $f(x) - p_n(x)$ could have any number.) In this case we simply select $n+2$ values where the error function has alternating minima and maxima, the remainder are discarded, the only proviso is that we must take the largest minima/maxima amongst the points we select. We will not give details of how to find the relevant minima/maxima but in general this is usually accomplished by a linear search along the error curve, not by differentiating. However, the location of suitable points by hand is a relatively simple procedure as the following example demonstrates.

Example 3.3.4 (Analytical)
Let us suppose that we have constructed a quadratic approximation for some function and the error curve looks like figure 3.3.5. The minimax approximation should have four alternating minima and maxima, this curve has five. We need to select four of the values shown in order to proceed with the Remes algorithm; in this case we take x_1, x_2, x_3 and x_4 as these are alternating minima and maxima and the maximum error is at x_4. The point x_0 will no longer be used.

As the order of the polynomial approximation increases the time required by the Remes algorithm increases rapidly and the closer the initial approximation is to the minimax the better. Furthermore, as the number of internal points increases equation (3.3.3) may become ill conditioned. The reason for this is clear. As n increases then the

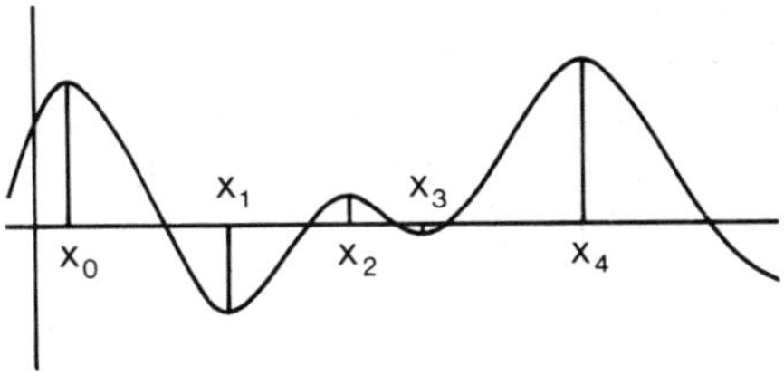

Figure 3.3.5 A possible quadratic approximation error curve.

matrix corresponding to equations (3.3.3) becomes very close to the Vandermonde matrix discussed in Chapter 2 which is known to be ill conditioned as the tabular points get closer together. A solution to this problem will be discussed in a later section.

Exercise 3.3.5 (Computational) Use the program **APPROX** to construct a minimax polynomial approximation for $\exp(x)$ which is of minimal order but has error no worse that 0.001 for $-1 \leq x \leq 1$.

Exercise 3.3.6 (Computational) Use the points -1, 0, 1 to construct a linear minimax approximation for x^3 over the interval $[-1, 1]$. Explain why the Remes algorithm fails in this case and suggest how it is possible to overcome this difficulty. (Hint: look at the quadratic minimax approximation.)

> 3.4 Chebyshev approximation

All the approximations we have discussed so far have been simple power series approximations, i.e. of the form

$$f(x) = \sum_{i=0}^{i=n} a_i x^i + a_{n+1} x^{n+1} + \ldots, \tag{3.4.1}$$

in the hope that, provided the series converges as $n \to \infty$, taking a finite number of terms will provide a suitable approximation. If we wish to obtain an approximation close to the minimax by truncating equation (3.4.1) after n terms then the remainder must satisfy the Chebyshev equi-oscillation property. Clearly, using this simple power series cannot produce this phenomenon, therefore, we will now look at alternative expansions. We begin by looking at Chebyshev polynomials.

Let us consider a sequence of functions which are defined, for all values of $x \in [-1, 1]$, as follows:

$$T_0(x) = 1,\ T_1(x) = x,$$

$$T_{n+1}(x) = 2xT_n(x) - T_{n-1}(x). \tag{3.4.2}$$

From the recurrence relation (3.4.2) we can generate

$$\begin{aligned} T_2(x) &= 2xT_1(x) - T_0(x) = 2x^2 - 1, \\ T_3(x) &= 2xT_2(x) - T_1(x) = 4x^3 - 3x, \end{aligned}$$

and so on. These functions have some remarkable properties. For example, notice that $T_n(x)$ is a polynomial of degree n and that for n odd T_n is an odd function whilst for n even T_n is even. The functions $T_n(x)$ are called *Chebyshev polynomials.* The first four are shown in figure 3.4.1.

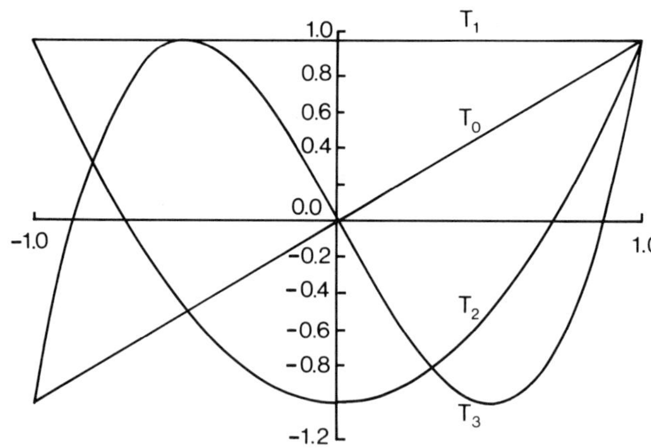

Figure 3.4.1 The Chebyshev polynomials $T_i(x), i = 0, 1, 2, 3$.

Exercise 3.4.1 (Analytical) Show that $T_n(x), x \in [-1, 1]$, is a polynomial of degree n and that it is odd when n is an odd integer and even when n is an even integer. (Hint: use induction.)

We can also define Chebyshev polynomials in another way, i.e.

$$T_n(x) = \cos(n \cos^{-1}(x)). \tag{3.4.3}$$

To see this note that

$$\begin{aligned} T_{n+1}(x) &= \cos((n+1) \cos^{-1}(x)) \\ &= \cos(\cos^{-1}(x)) \cos(n \cos^{-1}(x)) \\ &\qquad - \sin(\cos^{-1}(x)) \sin(n \cos^{-1}(x)) \end{aligned}$$

and

$$\begin{aligned} T_{n-1}(x) &= \cos((n-1) \cos^{-1}(x)) \\ &= \cos(\cos^{-1}(x)) \cos(n \cos^{-1}(x)) \\ &\qquad + \sin(\cos^{-1}(x)) \sin(n \cos^{-1}(x)) \end{aligned}$$

Adding together these equations gives

$$T_{n+1}(x) + T_{n-1}(x) = 2 \cos(\cos^{-1}(x)) \cos(n \cos^{-1}(x)) = 2xT_n(x), \tag{3.4.4}$$

which is equivalent to (3.4.2). The definition (3.4.3) now enables us to draw another conclusion about the polynomial $T_n(x)$: it has exactly $n+1$ alternating minima and maxima, with equal magnitude, in the interval $[-1, 1]$. The importance of this comment can be demonstrated as follows.

Let us assume that instead of (3.4.1) we can write

$$f(x) = \sum_{i=0}^{i=n} c_i T_i(x) + c_{n+1} T_{n+1} + c_{n+2} T_{n+2} + \ldots. \tag{3.4.5}$$

If this is the case then the dominant term in the error when this series is truncated after $n+1$ terms will be $c_{n+1}T_{n+1}$ and as T_{n+1} has exactly $n+2$ equal and opposite minima and maxima we should obtain an approximation close to the minimax! We must now justify the expansion (3.4.5) of any function in terms of Chebyshev polynomials. First of all, if the series (3.4.1) converges for $x \in [-1, 1]$ then so will any rearrangement of its terms, but this is just what the Chebyshev expansion is. Therefore, if (3.4.1) converges then so will (3.4.5). The next stage is to find an expression for the coefficients c_j, $j = 0, 1, \ldots$. This turns out to be quite straightforward because the Chebyshev polynomials possess an orthogonality property, i.e.

$$\int_{-1}^{1} \frac{T_i(x)T_j(x)}{\sqrt{(1-x^2)}}\,dx = \begin{cases} \pi, & i=j=0 \\ \frac{\pi}{2}, & i=j>0 \\ 0, & i\neq j \end{cases} .$$

To see this notice that the substitution $x = \cos\theta$ reduces

$$\int_{-1}^{1} \frac{T_i(x)T_j(x)}{\sqrt{(1-x^2)}}\,dx$$

to the standard trigonometric integral

$$\int_{0}^{\pi} \cos i\theta \cos j\theta \, d\theta = \begin{cases} \pi, & i=j=0 \\ \frac{\pi}{2}, & i=j>0 \\ 0, & i\neq j \end{cases} .$$

Therefore, in order to determine the coefficients c_j in equation (3.4.5) we multiply both sides by $T_j(x)/(1-x^2)^{1/2}$ and integrate over the interval $(-1,1)$, i.e.

$$c_0 = \frac{1}{\pi}\int_{-1}^{1} \frac{f(x)}{\sqrt{(1-x^2)}}\,dx = \frac{1}{\pi}\int_{0}^{\pi} f(\cos\theta)\,d\theta, \qquad (3.4.6)$$

$$c_j = \frac{2}{\pi}\int_{-1}^{1} \frac{T_j(x)f(x)}{\sqrt{(1-x^2)}}\,dx = \frac{2}{\pi}\int_{0}^{\pi} f(\cos(\theta))\cos j\theta\,d\theta. \qquad (3.4.7)$$

Alternatively, we can redefine the Chebyshev series (3.4.5) as

$$f(x) = \frac{1}{2}c_0 + \sum_{i=1}^{i=n} c_i T_i(x) + c_{n+1}T_{n+1} + c_{n+2}T_{n+2} + \ldots = \sum_{i=0}^{i=\infty}{}' c_i T_i(x), \qquad (3.4.8)$$

where the symbol $'$ denotes that the first coefficient is $\frac{1}{2}c_0$. All the coefficients are then given by (3.4.7).

Example 3.4.1 (Analytical)
Determine the Chebyshev coefficients for the function $3x(1-x^2)^{1/2}$. Notice that this is an odd function, i.e. $f(x) = -f(-x)$, and so all the even coefficients will vanish; we need only determine the odd coefficients. Therefore,

$$c_1 = \frac{2}{\pi}\int_{-1}^{1} \frac{3x\sqrt{(1-x^2)}T_1(x)}{\sqrt{(1-x^2)}}\,dx$$

$$= \frac{4}{\pi}\int_0^1 3x^2\,dx = \frac{4}{\pi},$$

$$c_3 = \frac{2}{\pi}\int_{-1}^1 \frac{3x\sqrt{(1-x^2)}\,T_3(x)}{\sqrt{(1-x^2)}}\,dx$$

$$= \frac{4}{\pi}\int_0^1 3x(4x^3-3x)\,dx = -\frac{12}{5\pi},$$

and so on. The function given by $c_1T_1(x) + c_3T_3(x)$ is compared with $f(x) = 3x(1-x^2)^{1/2}$ in figure 3.4.2 and the associated error is shown in figure 3.4.3. Notice that the error function has six alternating minima/maxima, but of unequal size. Although we have approximated f by a cubic, the six minima/maxima cause no problems because the coefficient c_4 is zero and therefore the quartic approximation for f is given by a cubic.

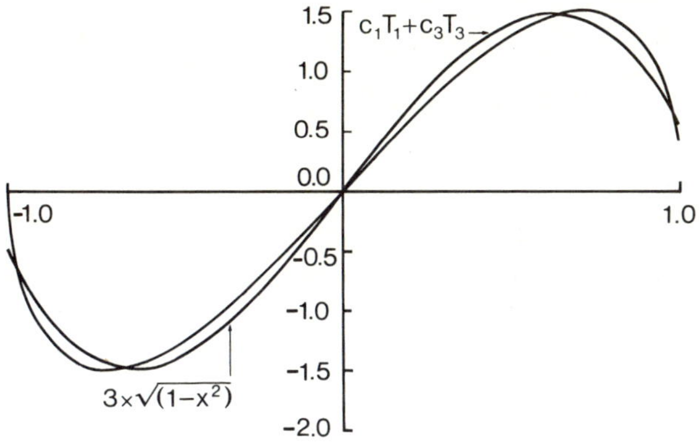

Figure 3.4.2 Cubic Chebyshev approximation for $3x(1-x^2)^{1/2}$.

Example 3.4.2 (Analytical)

In order to illustrate approximation using Chebyshev polynomials we shall compare a linear Chebyshev approximation for the function e^x, $|x| \leq 1$, with the corresponding linear minimax approximation $p_1(x) = 1.2464279 + 1.175201x$, which was determined in example 3.3.1. The coefficients for the required approximation are given, from equation (3.4.7),

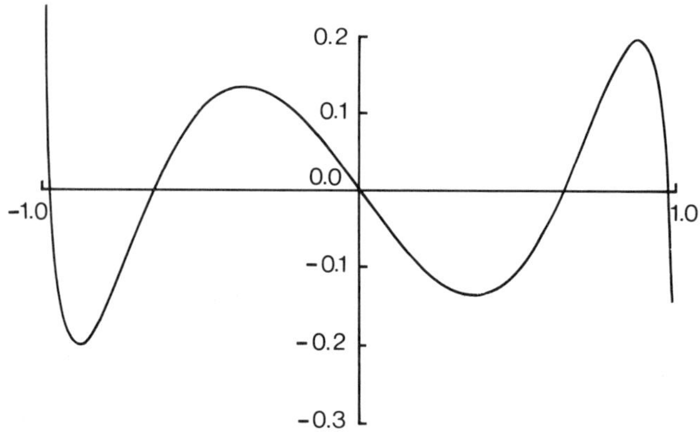

Figure 3.4.3 The error associated with figure 3.4.2.

as $c_0 = 2.53213212$ and $c_1 = 1.130318$, which gives the Chebyshev approximation

$$\exp(x) \approx 1.266066 + 1.130318T_1(x) = 1.266066 + 1.130318x$$

which is close to the minimax approximation of degree 1. (Notice that the first term is $\frac{1}{2}c_0$.)

Exercise 3.4.2 (Computational) Use the program **APPROX** with the option **Method: Chebyshev** to find approximations for the function $\exp(x)$, $x \in [-1, 1]$, of degree 1,2 and 3. Compare the computed approximations with the minimax approximations derived in the last section. (The resulting Chebyshev approximations are shown in figure 3.4.4.)

Given the coefficients of a Chebyshev approximation we can compare it with the minimax polynomial. For example, the minimax cubic approximation for $\exp(x)$, x in $[-1, 1]$, is given by

$$p_3(x) = 0.9945858 + 0.9956685x + 0.5429732x^2 + 0.1795327x^3$$

which has minimax error $\epsilon_3 = 0.0055216$. The corresponding Chebyshev approximation is

$$P_3(x) = 1.266066T_0 + 1.130318T_1 + 0.271495T_2 + 0.044337T_3$$

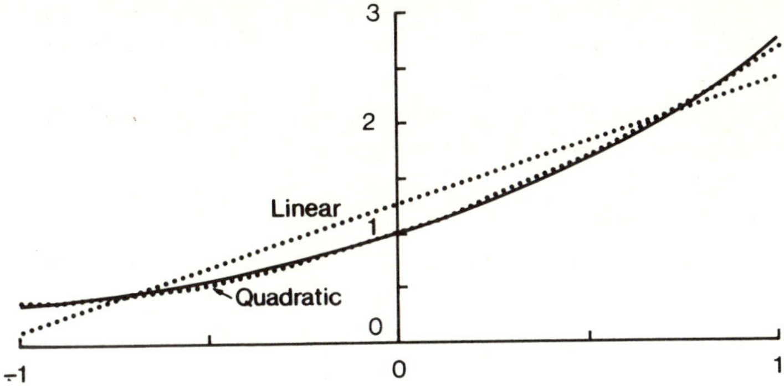

Figure 3.4.4 Chebyshev approximations for exp(x).

which gives the equivalent polynomial

$$P_3(x) = 0.994573 + 0.997307x + 0.542986x^2 + 0.177348x^3,$$

and this has maximum error 0.006065. At this point we emphasize that there is no need to convert a Chebyshev series in this way; in fact to do so may lead to significant errors. The reason for this is the size of the coefficients in the corresponding series; the smaller the coefficients the better. Notice that for this example the coefficients of the higher powers of the Chebyshev series are much smaller than those of the minimax approximation. It is better to evaluate a given Chebyshev series by making use of the recurrence relation (3.4.2) in order to minimize induced errors. In general given a Chebyshev series

$$f(x) = \sum_{i=0}^{i=n}{}' c_i T_i(x)$$

it can be evaluated as follows. For a given value of x define $u_n(x) = c_n$

$$\begin{aligned} u_{n-1}(x) &= c_{n-1} + 2xu_n, \\ u_j(x) &= c_j + 2xu_{j+1} - u_{j+2}, \qquad j = n-2, \ldots, 1, \end{aligned} \quad (3.4.9)$$

then

$$u_0(x) = \tfrac{1}{2}c_0 + xu_1(x) - u_2(x) = \sum_{i=0}^{i=n}{}' c_j T_j(x) = f(x).$$

Notice that (3.4.9) is a second-order recurrence relation, for any fixed value of x, and is relatively stable provided $|x| \leq 1$ which is precisely the range of values in which we are interested. (This technique for evaluating Chebyshev series can be compared with nested multiplication for power series.) In order to show this we rearrange the equations (3.4.9) to give

$$\begin{aligned} c_n &= u_n(x), \\ c_{n-1} &= u_{n-1}(x) - 2xu_n(x), \\ c_j &= u_j(x) - 2xu_{j+1}(x) + u_{j+2}(x), \qquad j = n-2, \ldots, 1, \\ \tfrac{1}{2}c_0 &= u_0(x) - xu_1(x) + u_2(x). \end{aligned}$$

Now recall that

$$\begin{aligned} f(x) &= c_nT_n + \ldots + c_1T_1 + \tfrac{1}{2}c_0 \\ &= u_n(x)(T_n(x) - 2xT_{n-1}(x) + T_{n-2}(x)) + \\ &\quad u_{n-1}(x)(T_{n-1}(x) - 2xT_{n-2}(x) + T_{n-3}(x)) + \ldots + \\ &\quad u_1(T_1(x) - xT_0(x)) + u_0(x)T_0(x). \end{aligned}$$

Finally, we use the recurrence relation (3.4.2), in which case all the terms except the last on the right-hand side vanish; this last term is u_0 since $T_0 = 1$. This technique would not be used to produce a explicit expression for a Chebyshev series as a power series but to evaluate the Chebyshev series at a particular point.

Example 3.4.3 (Calculator)
Evaluate the Chebyshev series

$$1.266066T_0 + 1.130318T_1 + 0.271495T_2 + 0.044337T_3$$

at the point $x = \frac{1}{2}$. We set $u_3 = 0.044337$ and then produce

$$\begin{aligned} u_2 &= 0.271493 + 2 \times \tfrac{1}{2} \times u_3 = 0.271493 + 0.044337 = 0.315832 \\ u_1 &= c_1 + 2 \times \tfrac{1}{2} \times u_2 - u_3 = 1.130318 + 0.315832 - 0.044337 \\ &= 1.401813 \\ u_0 &= \tfrac{1}{2}c_0 + \tfrac{1}{2}u_1 - u_2 = 1.6511405. \end{aligned}$$

(Recall that $\frac{1}{2}c_0 = 1.266066$ from equation (3.4.7).)

In general the determination of the Chebyshev coefficients is not as simple as demonstrated in example 3.4.1 and it may be necessary to resort to numerical integration in order to evaluate them. (This is what the option **Chebyshev** in the program **APPROX** does.) However, the Chebyshev expansion (3.4.5) has one other bonus. Clearly, given any other set of polynomials we could expand any suitable function in terms of a similar expansion in which case we would still need to determine the relevant coefficients. Amongst all[1] such expansions it can be shown that the coefficients in the Chebyshev series decrease more rapidly than any other expansion. Furthermore, given an expansion in terms of the first n such polynomials the error in truncating the series after n terms, i.e.

$$c_nT_n + c_{n+1}T_{n+1} + \dots$$

is closely bounded above by the single term $|c_nT_n|$, provided the coefficients tend to zero sufficiently rapidly. (See Fox and Parker (1972).) However, $|T_n(x)| \leq 1$ and so we can bound the truncation error by the single term $|c_n|$. Therefore, if we require an approximation which has an overall error bound ϵ we determine successive Chebyshev coefficients until we have $|c_n| \leq \epsilon$ and this will give the required polynomial.

Example 3.4.4 (Computational)
The cubic Chebyshev approximation for $\exp(x)$, $x \in [-1, 1]$ is given by

$$P_3(x) = 1.266066T_0 + 1.130318T_1 + 0.271493T_2 + 0.044337T_3$$

where the next coefficient is given by $c_4 = 0.005474$. The corresponding cubic minimax is given by

$$p_3(x) = 0.9945858 + 0.9956685x + 0.5429732x^2 + 0.1795327x^3,$$

which has minimax error $\epsilon_3 = 0.0055216$. From this we can see that c_4 is a reasonable estimate for ϵ_3.

In many cases Chebyshev approximations will be very close to the minimax polynomial. To illustrate this we list in table 3.4.1 various functions and compare the minimax error, corresponding Chebyshev maximum error and also the size of the first coefficient ignored in the Chebyshev approximation[2]. If the Chebyshev approximation is not sufficiently close to the minimax approximation then we simply take it as a

[1] Strictly speaking amongst all expansions in terms of ultra-spherical polynomials for which the associated weight function is $w(x) = (1 - x)^{-p}$, $p = \frac{1}{2}$ gives Chebyshev polynomials, $p = 0$ gives Legendre polynomials.

[2] The values given are those calculated by the software. If exact coefficients are supplied then different entries may be obtained.

Table 3.4.1 Minimax and Chebyshev maximum absolute errors.

	Order	Taylor series error	Minimax error	Chebyshev error	Coefficient neglected
	3	0.04744	0.005529	0.006063	0.005474
$\exp(x)$	4	0.00656	0.000547	0.000587	0.000522
$x \in [-1, 1]$	5	0.00292	0.000045	0.000064	0.000046
	2	0.28737	0.029830	0.033308	-0.029437
$\log_e(1 + x^2)$	4	0.15340	0.003424	0.003887	0.003198
$x \in [-1, 1]$	6	0.07556	0.000444	0.000689	-0.000432
$\sin(x)$	3	0.00897	0.000499	0.000503	0.000521
$x \in [-1, 1]$	5	0.00009	0.000003	0.000003	-0.000003

first approximation in the Remes algorithm. Alternatively, we can start the Remes algorithm by constructing a polynomial approximation using the points where the maximum errors occur at the extrema of the required Chebyshev polynomial. For example, if we require a polynomial approximation of degree n, then we know that the error function must have $n + 2$ alternating minima/maxima; we take as our first approximation for the extrema the points $x_i = \cos\left(\frac{i\pi}{(n+1)}\right)$, $i = 0, 1, \ldots, n + 1$, which are the $n + 2$ extrema of $T_{n+1}(x)$. This choice of points should produce an approximation which is close to the minimax.

Exercise 3.4.3 (Analytical) As an alternative to the Chebyshev polynomials, given on page 68, we can define the Modified Chebyshev polynomials $T_n^*(x)$, for $0 \leq x \leq 1$, according to

$$T_n^*(x) = T_n(2x - 1),\ 0 \leq x \leq 1.$$

Derive the first five Modified Chebyshev polynomials and sketch the corresponding curves. Show that

$$T_{n+1}^*(x) = 2(2x - 1)T_n^*(x) - T_{n-1}^*(x).$$

The best quadratic Chebyshev approximation for $x^3 + x^2$ over the range $[-1, 1]$ is $x^2 + \frac{3}{4}x$ which has a maximum error 0.25, i.e.

$$x^3 + x^2 = x^2 + \tfrac{3}{4}x + \tfrac{1}{4}T_3(x).$$

Show that the best Chebyshev quadratic approximation for this function over the range [0, 1] is given by

$$x^3 + x^2 = \tfrac{1}{32}(10T_0^* + 15T_1^* + 6T_2^*) + x^2 + \tfrac{1}{32}T_3^*.$$

Therefore, the best quadratic approximation is $\frac{1}{32}(1 - 18x + 80x^2)$ which has a maximum error of $\frac{1}{32}$, one eighth that of the approximation on the interval $[-1, 1]$.

Exercise 3.4.4 (Analytical). Runge's problem revisited. In example 2.1.3 and exercise 2.1.4 the construction of a polynomial approximation for $\exp(x)$ was considered. It was demonstrated that if the interpolation points are equally spaced then as the number of points increases the interpolating polynomial, $p_n(x)$, oscillates wildly. This would seem to imply that

$$\max_{x_0 \le x \le x_n} |e_n(x)| = \max_{x_0 \le x \le x_n} |\exp(-x^2) - p_n(x)| \to \infty \text{ as } n \to \infty.$$

However, recall that in general the error between an interpolating polynomial $p_n(x)$ and a function, $f(x)$, for which it is an approximation satisfies

$$|f(x) - p_n(x)| \le \frac{M_{n+1}}{(n+1)!} \prod_{i=0}^{i=n} (x - x_i),$$

where $M_n = \max_{x_0 \le x \le x_n} |f^{(n)}(x)|$ and x_i, $i = 0, \ldots, n$, are the interpolation points selected. Show that by selecting

$$\prod_{i=0}^{i=n} (x - x_i) = 2^{-n} T_{n+1}(x) \tag{3.4.10}$$

the error function $e_n(x)$ satisfies

$$|e_n(x)| \le \frac{M_{n+1}}{2^n (n+1)!}.$$

Therefore, provided M_{n+1} is finite for all n, $e_n(x) \to 0$ as $n \to \infty$. The choice (3.4.10) looks rather strange but simply requires that the zeros of both sides of this equation are identical, i.e. we need to select the interpolation points to be the zeros of T_{n+1} which are

$$x_i = \cos\left(\frac{(2i+1)\pi}{2n+2}\right), \qquad i = 0, 1, \ldots, n. \tag{3.4.11}$$

> 3.4.1 Polynomial economization

Let us suppose that we are given a polynomial of degree n and ask is it possible to construct a polynomial of degree $\leq n$ which is minimax? This apparently difficult problem is in fact very easy to solve using Chebyshev polynomials. In order to see this let

$$Q_n(x) = a_0 + a_1 x + \ldots + a_{n-1}x^{n-1} + a_n x^n,$$

and then note that $T_n(x) = 2^{n-1}x^n + P_{n-1}(x)$, where P_{n-1} is a polynomial of degree $n-1$. Therefore,

$$x^n = 2^{(1-n)}(T_n(x) - P_{n-1}(x))$$

which gives

$$\begin{aligned} Q_n(x) &= a_0 + a_1 x + \ldots + a_{n-1}x^{n-1} + a_n 2^{(1-n)}(T_n(x) - P_{n-1}(x)) \\ &= a_0 + a_1 x + \ldots + a_{n-1}x^{n-1} - a_n 2^{(1-n)} P_{n-1}(x) + a_n 2^{(1-n)} T_n(x). \end{aligned}$$

Therefore, the $(n-1)$th degree polynomial

$$a_0 + a_1 x + \ldots + a_{n-1}x^{n-1} - a_n 2^{(1-n)} P_{n-1}(x)$$

differs from Q_n by a term $a_n 2^{(1-n)} T_n(x)$ which has exactly $n+1$ equal and opposite minima/maxima of size $a_n 2^{(1-n)}$ and is therefore the minimax approximation of this degree.

Example 3.4.5 (Analytical)
Construct a quadratic minimax approximation for the polynomial

$$Q_3(x) = 1 + x + x^2 + x^3.$$

The Chebyshev polynomial $T_3(x) = 4x^3 - 3x$, therefore,

$$x^3 = \tfrac{1}{4}(T_3(x) + 3x)$$

and

$$\begin{aligned} Q_3 = 1 + x + x^2 + x^3 &= 1 + x + x^2 + \tfrac{1}{4}(T_3(x) + 3x) \\ &= 1 + \tfrac{7}{4}x + x^2 + \tfrac{1}{4}T_3(x). \end{aligned}$$

Therefore, the function $1 + \frac{7}{4}x + x^2$ is the required quadratic minimax approximation for $Q_3(x)$ since it will have precisely four equal and opposite minima/maxima of size $\frac{1}{4}$. No other quadratic polynomial can have a smaller error.

Exercise 3.4.5 (Analytical) Construct a linear minimax approximation for the quadratic function

$$Q_2(x) = 1 + x + \tfrac{1}{2}x^2$$

which is the Taylor series expansion of $\exp(x)$ about $x = 0$. Determine the minimax approximation error involved.

> 3.4.2 Chebyshev minimax approximation

In section 3.3 the Remes algorithm was used to construct the minimax approximation for a given function. This algorithm is relatively easy to implement as was demonstrated in exercise 3.3.2 but as the order of the approximating polynomial increases convergence may be slow and it is possible that the process becomes ill conditioned. In the previous section it was shown how to construct a Chebyshev approximation which is almost as good as the minimax and so it would seem sensible to combine the Remes algorithm together with Chebyshev polynomials; we will briefly investigate this further in this section.

Let $f(x)$ be a continuous function for which $p_n(x)$ is the minimax polynomial of degree n. The error function $e_n(x) = f(x) - p_n(x)$ has exactly $n+2$ alternating minima and maxima of equal magnitude. However, as $p_n(x)$ is a polynomial of degree n we can express it in the form

$$p_n(x)+ = \tfrac{1}{2}c_0T_0 + c_1T_1(x) + c_2T_2(x) + \ldots + c_nT_n(x).$$

This form of expressing the minimax polynomial has several advantages over an expansion in simple powers of x, not the least of which is that the coefficients will be numerically smaller which means that the evaluation of this expression will be less prone to rounding errors. Any other linear combination of Chebyshev polynomials will have a larger maximum error bound than that given by the minimax error $\epsilon_n = \max|e_n(x)|$. Let us assume that we have obtained an approximation for the minimax polynomial in terms of a Chebyshev series and that we have determined $(n+2)$ points at which the associated error function has alternating minima and maxima, $x_0 \leq x_1 \leq \ldots \leq x_{n+1}$. Now we determine a better polynomial approximation, $q_n(x)$, as

$$q_n(x) = \tfrac{1}{2}b_0 + b_1T_1(x) + \ldots + b_nT_n(x)$$

by solving the $(n+2)$ *linear* equations

$$f(x_i) - q_n(x_i) = (-1)^{i+1}E, \qquad i = 0, \ldots, n+1, \tag{3.4.12}$$

to determine the coefficients b_0, b_1, ..., b_n. The similarity between the equations (3.4.12) and (3.3.3) is striking but the former equations have the advantage of being far better conditioned as n increases. In order to demonstrate this procedure consider example 3.4.6.

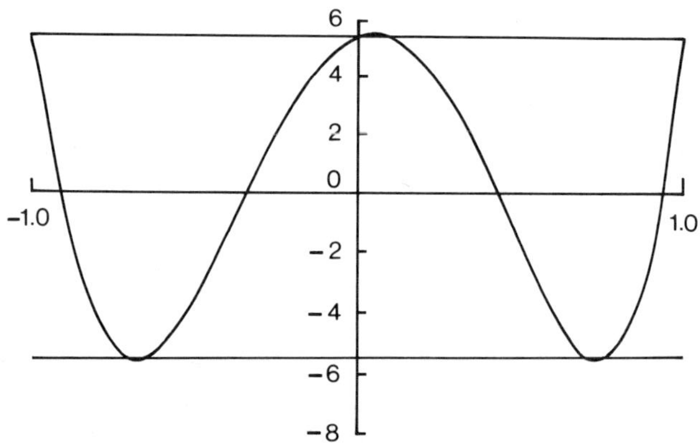

Figure 3.4.5 The error $\exp(x) - Q_3(x)$. See equation (3.4.12).

Example 3.4.6 (Computational)

Let us construct the cubic Chebyshev minimax approximation for the function $\exp(x), -1 \leq x \leq 1$. We begin by specifying five points as extrema for the error function

$$e_3(x) = \exp(x) - c_0T_0 - c_1T_1 - c_2T_2 - c_3T_3.$$

We use the extrema of the Chebyshev polynomial $T_4(x)$ which are given by $x_i = \cos\left(\frac{1}{4}i\pi\right)$, $i = 0, 1, 2, 3, 4$. (See page 76.) This gives the points $x_0 = -1$, $x_1 = -\frac{1}{2}\sqrt{2}$, $x_2 = 0$, $x_3 = \frac{1}{2}\sqrt{2}$ and $x_4 = 1$. (Compare these values with those used in example 3.3.3). We now solve the five linear equations

$$\begin{aligned} \exp(x_i) - c_0T_0(x_i) - c_1T_1(x_i) - c_2T_2(x_i) - c_3T_3(x_i) &= (-1)^i E, \\ i &= 0, 1, 2, 3, 4. \end{aligned}$$

These equations can be rewritten as

$$\begin{pmatrix} 1 & -1 & 1 & -1 & 1 \\ 1 & \frac{1}{2}\sqrt{2} & 0 & -\frac{1}{2}\sqrt{2} & -1 \\ 1 & 0 & -1 & 0 & 1 \\ 1 & -\frac{1}{2}\sqrt{2} & 0 & \frac{1}{2}\sqrt{2} & -1 \\ 1 & 1 & 1 & 1 & 1 \end{pmatrix} \begin{pmatrix} c_0 \\ c_1 \\ c_2 \\ c_3 \\ E \end{pmatrix} = \begin{pmatrix} \exp(-1) \\ \exp(-\frac{1}{2}\sqrt{2}) \\ \exp(0) \\ \exp(\frac{1}{2}\sqrt{2}) \\ \exp(1) \end{pmatrix}$$

and produce the cubic Chebyshev polynomial

$$Q_3(x) = 1.266066 + 1.130321T_1 + 0.271540T_2 + 0.044880T_3 \quad (3.4.13)$$

which has the error function shown in figure 3.4.5. This function has maximum absolute value 0.00583 but the corresponding value of E is 0.005474. We see that this cubic is not minimax but we can repeat this process to improve it. We can also compare this cubic with the first four terms of the Chebyshev series for $\exp(x)$ given by

$$\exp(x) \approx 1.266066 + 1.130318T_1 + 0.271495T_2 + 0.044337T_3$$

which has a maximum error of 0.00607.

Exercise 3.4.6 (Computational) Write out the equations in the next step of the Remes algorithm for determining the cubic Chebyshev minimax approximation for $\exp(x)$. Solve the corresponding linear equations and investigate whether the resulting cubic is minimax.

> 3.5 Orthogonal functions

The Chebyshev polynomials introduced in section 3.4 are extremely useful when an approximation to a given function is required. This is because they possess many useful properties, one of the most important of these being orthogonality. By this we mean that for a set of polynomial functions, $\phi_n(x)$, $-1 \le x \le 1$, $n = 0, 1, \ldots$, there exists a weight function $w(x) > 0$ such that

$$\int_{-1}^{1} w(x)\phi_n(x)\phi_m(x)\,dx = M_n\delta_{mn},$$

where M_n is a constant and δ_{mn} is 1 if $n = m$ and 0 otherwise. We could take any other set of linearly independent polynomials, $\phi_n(x)$, $x \in [-1, 1]$, and consider expansions similar to (3.4.1) or (3.4.5). (The condition of linear independence is needed to ensure that no one member of the set can be expressed in terms of a linear combination of the rest and would, therefore, be redundant.) Ignoring the problems of convergence of infinite series, let us for the moment write

$$f(x) = c_0\phi_0 + c_1\phi_1 + c_2\phi_2 + \dots .$$

However, if the functions ϕ_0, ϕ_1, ...are orthogonal we can determine the coefficients c_n as

$$c_n = \frac{1}{M_n}\int_{-1}^{1} w(x)f(x)\phi_n(x)\,dx.$$

Example 3.5.1
The following are sets of orthogonal polynomials

(i) *Chebyshev polynomials.* $T_n(x)$, $x \in [-1,1]$, $w(x) = (1-x^2)^{-1/2}$.

$$T_0(x) = 1,\ T_1(x) = x,$$

$$T_{n+1}(x) = 2xT_n(x) - T_{n-1}(x).$$

(ii) *Legendre polynomials.* $P_n(x)$, $x \in [-1,1]$, $w(x) = 1$.

$$P_0(x) = 1, P_1(x) = x,$$

$$(n+1)P_{n+1}(x) + nP_{n-1}(x) = (2n+1)xP_n(x).$$

Further details of these polynomial functions and their properties can be found in Fox and Parker (1972) or Churchill (1963).

The idea of orthogonal polynomials can be generalized further to sets of orthogonal functions.

Example 3.5.2
The following are sets of orthogonal functions

(i) Exponential functions $\exp(inx)$, $i = \sqrt{(-1)}$, $w(x) = 1$.

(ii) *Fourier series.*

$$\phi_n(x) = \begin{cases} \sin(n\pi x) \\ \cos(n\pi x) \end{cases},\ w(x) = 1.$$

Exercise 3.5.1 (Analytical) Show that the sets of functions given in examples 3.5.1 and 3.5.2 are orthogonal. In each case, assuming that there exists an expansion of the form,

$$f(x) = \sum_{n=0}^{n=\infty} c_n\phi_n(x), \tag{3.5.1}$$

determine an expression for the coefficients c_n, $n = 0, 1, \dots$

Whenever we express a function in terms of an infinite series such as (3.5.1) we must always consider whether the series representation is valid, i.e. does the series converge and if so does it converge to the function we require. If the function is continuous then there are few problems, however, even when the function is discontinuous it is sometimes possible to determine a suitable expansion. We will consider this problem in the next section.

> 3.6 Fourier series expansions

The idea of orthogonal functions was briefly introduced in the last section, we can demonstrate how useful this notion can be. Consider the following expansion in terms of the orthogonal functions $\cos(n\pi x)$, i.e.

$$\begin{aligned} f(x) = \cos(2\pi x) &+ \tfrac{1}{9}\cos(6\pi x) + \tfrac{1}{25}\cos(10\pi x) + \dots \\ &+ \frac{1}{(2n-1)^2}\cos(2(2n-1)\pi x) + \dots . \end{aligned} \tag{3.6.1}$$

For the moment let us ignore the problem of whether this infinite series converges and try instead to plot it. Of course we cannot add in all the terms but for the moment let us see what happens if we include the first four, five or perhaps six terms. Now we have a finite expression which we can plot. The result is shown in figure 3.6.1.

At first sight it appears that as we include more and more terms the sum of this series of continuous functions appears to be a function which is not differentiable at various points. Conversely, we seem to be able to write the sawtooth waveform as a sum of eminently smooth functions. We can now ask whether this is always the case. However, before we answer this question let us take a closer look at the graph drawn in figure 3.6.1, in particular near the peaks. By plotting in more detail we see that the graph in fact does not change direction sharply but does so in a smooth way (see figure 3.6.2). However as we include more and more terms the magnification required to see this smooth change becomes greater and greater and to all practical purposes the sawtooth wave is generated.

Let us now turn to a general function $f(x)$, $|x| \leq 1$, and examine the conditions under which it is possible to write this function as an

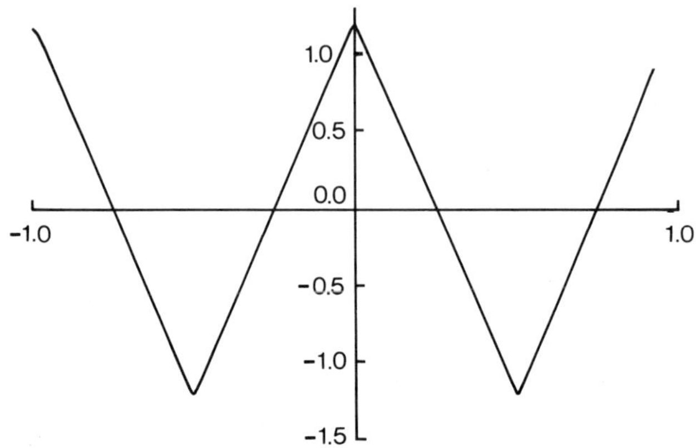

Figure 3.6.1 The cosine series (3.6.1).

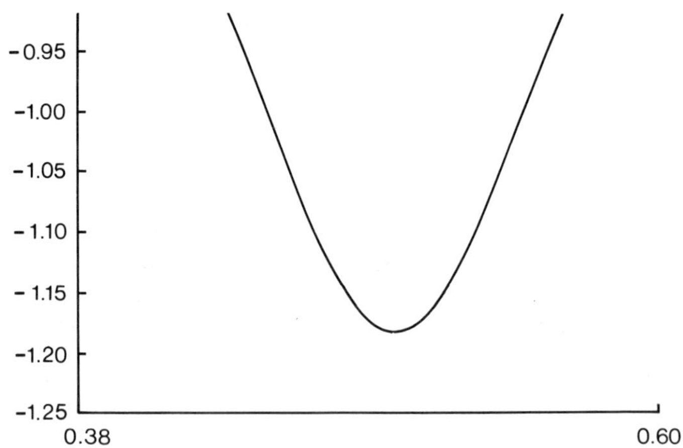

Figure 3.6.2 Enlargement of figure 3.6.1 near $x = 0.5$, $y = -1.23$.

expansion of the form

$$f(x) = \frac{a_0}{2} + \sum_{n=1}^{n=\infty} \left(a_n \cos(n\pi x) + b_n \sin(n\pi x) \right), \quad |x| \leq 1, \qquad (3.6.2)$$

where the $'$ denotes that the first coefficient is given by $\frac{1}{2}a_0$. The expansion (3.6.2) is called the Fourier series expansion of $f(x)$. If we are given a function $f(x)$ for $a \leq x \leq b$ then we can always map the function to the interval $|x| \leq 1$ by a suitable linear transformation. Alternatively, if the function is defined on an infinite interval then since the sine and cosine functions are periodic, i.e. $\sin(x+2\pi) = \sin(x)$ and $\cos(x+2\pi) = \cos(x)$, we shall require that the function $f(x)$ is also periodic. Therefore, without loss of generality we shall only consider functions which are defined for $|x| \leq 1$. Next we consider the notion of sectional continuity.

Definition 3.6.1. Let $f(x)$ be a function which is continuous at all points of a finite interval $a \leq x \leq b$ except possibly at a set of points $a \leq x_1 < \ldots < x_p \leq b$. If f has a finite limit as x approaches the ends of each interval (x_i, x_{i+1}), $i = 1, \ldots, p-1$, from its interior then f is said to be sectionally continuous.

Definition 3.6.2. A function is said to have a right- (left-) hand derivative at a point if

$$\lim_{h \to 0} \frac{f(x+h) - f(x)}{h}$$

exists, where h tends to zero through negative (positive) values.

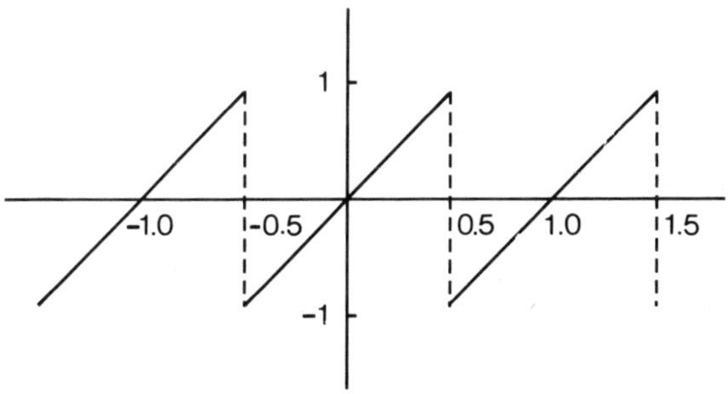

Figure 3.6.3 Sectionally continuous functions.

Example 3.6.1
The functions shown in figure 3.6.3 are simple examples of sectionally continuous functions.

Theorem 3.6.1. Let f be a sectionally continuous function on the interval $(-1, 1)$ then the Fourier series

$$\tfrac{1}{2}a_0 + \sum_{n=1}^{n=\infty} (a_n \cos(n\pi x) + b_n \sin(n\pi x)),$$

where

$$\begin{aligned} a_n &= \int_{-1}^{1} f(x)\cos(n\pi x)\,dx, \quad n = 0, 1, 2, \ldots \\ b_n &= \int_{-1}^{1} f(x)\sin(n\pi x)\,dx, \quad n = 1, 2, \ldots \end{aligned}$$

converges to the value

$$\tfrac{1}{2}[f(x+0) + f(x-0)]$$

at every point where f has a right- and left-hand derivative. (By $f(x \pm 0)$ we shall mean the limit of f at x through values which approach x from above or below.)

Example 3.6.2 (Computational)
Load the program **APPROX** and select the **Method: Fourier**. We will now investigate the Fourier series which correspond to the function

$$f(x) = \begin{cases} 1, & -1 < x < 0, \\ 0, & 0 < x < 1. \end{cases}$$

The function editor can be used to input this function piece by piece. To return to the function editor press ESCAPE and then when the prompt shows **Action: Continue** press RETURN which should produce the option **Function: as set**. (Notice that the current function and range is shown in the bottom window.) By pressing any key except RETURN cycle through the options available at this time until it reads **Function: inspect/edit** and then press RETURN again. This will display another prompt which you will need to alter to **Change f(x)** before pressing RETURN. A flashing cursor will appear in the bottom window and we can type in the value 1 which now becomes the current function. The next stage is to split the range using the prompt **Split section**, recall that we need to split the range at $x = 0$. The bottom window will now show that we have the function $f(x) = 1$ for negative values of x as the first of two sections. Next we shall amend the second section using the prompt **Next section** and then cycling until the prompt reads **Change f(x)**, in which case we can set the function to $f(x) = 0$. Finally, we can plot the current function using the option **Plot function**. (There are several other options available at this level which can be used to move between the various sections of the function, edit or delete sections, or add other sections if required.) At this point the required function should be plotted; notice in particular the form of the curve near the discontinuous point $x = 0$. This is caused by using only a finite number of points when plotting. Now quit the function editor and return to the prompt **Method: Fourier** and then press RETURN again. Again at this point there are various options available which include supplying the coefficients manually, reading them from file, storing the current coefficients in a file and calculating the coefficients corresponding to the current function. This last option requires the evaluation of equation (3.6.2) and in general this is carried out numerically. Therefore, accept the option **calculate coeffs** and then set the upper limit on the number of terms to four. After a short pause, whilst the coefficients are being calculated, the option will return to **Option: calculate coeffs**. Again cycle through the possibilities available at this point and plot the function, approximation and error. Increase the number of terms included in the series and compare the approximations which are determined. Notice that each approximation takes the value $\frac{1}{2}$ at the origin which agrees with the value given by theorem 3.6.1. How do you account for the behaviour of the Fourier series at ± 1?

Exercise 3.6.1 (Analytical) Determine analytically the Fourier coefficients for the function given in example 3.6.2 and compare them with those determined by the program **APPROX**.

Exercise 3.6.2 (Analytical) Show that the function $f(x) = \sqrt{|x|}$, $|x| < 1$ does not have a one-sided derivative at the point $x = 0$, but that this function does have a convergent Fourier series on the interval $|x| < 1$ which is valid at $x = 0$. This demonstrates that the existence of one-sided derivatives is not a necessary condition for the existence and convergence of Fourier series.

Exercise 3.6.3 (Computational/Analytical) Use **APPROX** to determine the Fourier series expansion of $\exp(x)$, $|x| \leq 1$. Compare the result with corresponding polynomial approximations. Investigate the behaviour of the Fourier series which is obtained for values $|x| > 1$.

Exercise 3.6.4 (Computational) Determine the Fourier series associated with the function $|\sin(\pi x)|$, $|x| \leq 1$.

Exercise 3.6.5 (Computational) Load the program **APPROX** and then input the file **RAMP**. Plot the function which is specified by this file and then determine its Fourier series. Compare the coefficients obtained with the exact analytical results. Compare the function and its Fourier approximation as the number of harmonics increases.

> 3.7 Rational approximation

In previous sections we have seen how it is possible to determine approximations for continuous functions in terms of polynomial expansions and sectionally continuous periodic functions in terms of Fourier series. However, if we are given a function which has an apparent singularity neither of these methods can give a satisfactory approximation. Consider the following example.

Example 3.7.1 (Computational)
Load the program **APPROX** and determine a cubic approximation for the function $f(x) = (x^2 - 3x + 2)^{-1}$, $|x| \leq 1.5$. Using the interpolation points -1.5, -1, 0, 1.5 produces the cubic shown in figure 3.7.1. The poor accuracy is due to the behaviour of f near $x = 1$. This demonstrates that attempting to approximate a function which has a singularity by continuous functions cannot lead to a satisfactory result. Let us reconsider the function $f(x) = (x^2 - 3x + 2)^{-1}$ and set $F(x) = (x-1)f(x)$. If we now determine a polynomial approximation for $F(x)$ then we obtain a very satisfactory minimax approximation. This suggests that the correct way of approximating the function $f(x)$ is to seek approximations of the form

$$f(x) \approx \frac{P_n(x)}{Q_m(x)},$$

where $P_n(x)$ and $Q_m(x)$ are polynomials of degree n and m respectively. Clearly, we could multiply P_n and Q_m by any factor and so by convention we assume that the coefficient of x^m in Q_m is 1. The resulting approximation is denoted by $R_{nm}(x)$.

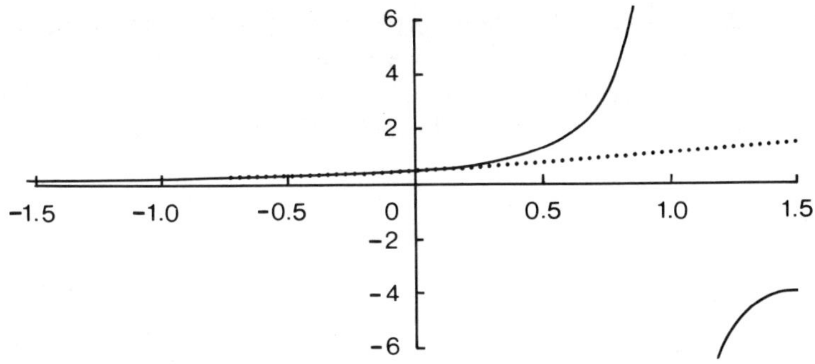

Figure 3.7.1 Cubic approximation for $(x^2 - 3x + 2)^{-1}$, $|x| \leq 1.5$.

Example 3.7.2 (Analytical)
In order to illustrate rational approximation let us determine $R_{1,1}(x)$ for the function $\exp(x)$, $|x| \leq 1$. Let the rational approximation $R_{1,1}$ be

$$R_{1,1}(x) = \frac{a_0 + a_1 x}{b_0 + x}$$

then we need to determine values for a_0, a_1 and b_0. As with polynomial interpolation we can only make $R_{1,1}$ and $\exp(x)$ agree at a finite set of points and since we have three coefficients to select we ensure that

$$\exp(x_i) = R_{1,1}(x_i),\ i = 1, 2, 3.$$

For example, let us use the tabulation points $x_1 = -1$, $x_2 = 0$ and $x_3 = 1$; then

$$\begin{aligned} &\text{at } x = -1 \quad &:& \quad (b_0 - 1)\exp(-1) = a_0 - a_1, \\ &\text{at } x = 0 \quad &:& \quad b_0 = a_0, \\ &\text{at } x = 1 \quad &:& \quad (b_0 + 1)\exp(1) = a_0 + a_1. \end{aligned}$$

Solving these equations produces $a_1 = -1$, $a_0 = b_0 = -2.1639534$. This gives the following rational approximation for $\exp(x)$:

$$R_{1,1}(x) = \frac{-2.1639534 - x}{-2.1639534 + x}$$

which is shown in figure 3.7.2. Notice that this function is singular at $x = -2.1639534$ but this is outside the region where the approximation is required. The error function corresponding to this approximation, $e_{1,1}(x) = \exp(x) - R_{1,1}(x)$, is shown in figure 3.7.3 and has an error bound of 0.0575. This error bound can be compared with the linear minimax error bound of 0.2788 and cubic minimax error bound of 0.0552.

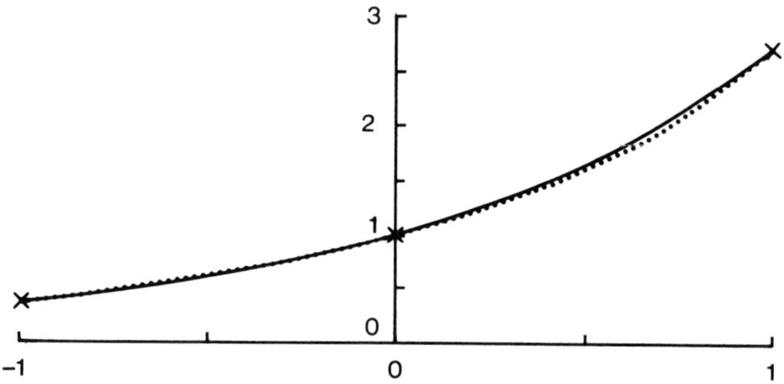

Figure 3.7.2 Rational approximation $R_{1,1}(x)$, for $\exp(x)$, $|x| \leq 1$.

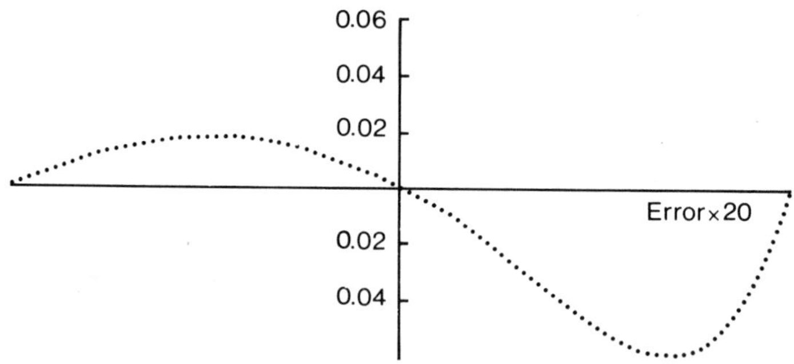

Figure 3.7.3 The error $R_{1,1}(x) - \exp(x)$, $|x| \leq 1$.

In general if we seek a rational approximation, $R_{nm}(x)$, for a given function $f(x)$ then we can match the function and the approximation at $n+m+1$ points since this is the number of coefficients to be determined, i.e. we select points x_i, $i = 0, \dots, (n+m)$, and then solve

$$Q_m(x_i)f(x_i) = P_n(x_i), \quad i = 0, \dots, (n+m).$$

Since the coefficient of x^m in Q_m is unity we can rewrite these equations as

$$\sum_{j=0}^{j=n} a_j x_i^j - \sum_{j=0}^{j=m-1} b_j x_i^j f(x_i) = x_i^m \, f(x_i),\ i = 0, \dots, (n+m). \qquad (3.7.1)$$

This gives a system of $(n+m+1)$ linear equations in the unknown coefficients which can be solved by Guassian elimination with partial pivoting, if necessary.

Exercise 3.7.1 (Computational) Use the program **APPROX** with the option **Method: Rational** to determine the $R_{2,2}(x)$ approximation for $\exp(x)$, $|x| \le 1$, at equally spaced points. Compare the results obtained by using other points. Also compare the results with $R_{3,1}$ and $R_{1,3}$ approximations for this function.

Exercise 3.7.2 (Computational) Use the program **APPROX** to construct rational approximations, $R_{mn}(x)$, for $\exp(x)$, $|x| \le 2$, for which m and n are less than 4. Compare them with the cubic minimax approximation.

As with polynomial approximation of functions based on interpolation at a set of points, rational approximation can suffer from induced ill conditioning. This can be overcome to some extent by using rational Chebyshev approximations and further details are given by Fox and Parker (1972). However, there are other problems which can occur.

(i) The equations (3.7.1) may turn out to be singular.

Example 3.7.3 (Analytical)
Let us consider a rational approximation, $R_{1,1}(x)$, for the cubic function

$$f(x) = x^3 - 3x^2 + 4x,$$

which is determined by interpolation at $x = 0$, $x = 1$ and $x = 2$. The equations (3.7.1) become

$$\begin{pmatrix} 1 & 0 & 0 \\ 1 & 1 & -2 \\ 1 & 2 & -4 \end{pmatrix} \begin{pmatrix} a_0 \\ a_1 \\ b_0 \end{pmatrix} = \begin{pmatrix} 0 \\ 2 \\ 8 \end{pmatrix}$$

and these equations are inconsistent.

(ii) In example 3.7.1 we saw that the rational approximation $R_{1,1}$ for $\exp(x)$, $|x| \leq 1$, determined using interpolation at -1,0 and 1 is

$$R_{1,1}(x) = \frac{-2.1639534 - x}{-2.1639534 + x}$$

which has a singularity at $x = 2.1639534$. Fortunately, this singularity is outside the region where the approximation is required, but this may not always be the case.

Example 3.7.4 (Analytical)
Consider the function

$$f(x) = -3 + 20x - 8x^2, \quad 0 \leq x \leq 2.$$

Using the tabulation points 0, 1 and 2 produces the rational approximation

$$R_{1,1} = \frac{1.5 + 3x}{-0.5 + x}$$

which has a singularity at $x = \frac{1}{2}$ where the original function is continuous.

(iii) As P_n and Q_m are polynomials they may have a common factor which can be cancelled out.

Example 3.7.5 (Analytical)
Consider the cubic

$$f(x) = 0.5x^3 - 4x^2 + 10.5x - 6$$

for which

$$f(1) = 1, f(2) = 3, f(3) = 3, f(4) = 4.$$

The rational function $R_{2,1}$ determined from equation (3.7.1) is

$$\begin{aligned} R_{2,1}(x) &= \frac{x^2 - 2x}{x - 2} = \frac{x(x-2)}{x-2} \\ &= x, \text{ if } x \neq 2. \end{aligned}$$

In order that the resulting function is continuous at $x = 2$ we need to define $R_{2,1}(2) = 2$ which is inconsistent with the data provided.

We have seen that it is posible that rational approximation may introduce singularities, but this possibility can be advantageous as the following simple exercise demonstrates.

Exercise 3.7.3 (Computational) The function $f(x) = \exp(x)/(1-2\cos(x))$ has a singularity at $x = \frac{1}{3}\pi$. Construct a rational approximation for $f(x)$ by considering the function $(x - \frac{1}{3}\pi)f(x)$.

> 3.7.1 Rational minimax approximation

Clearly, given a function $f(x)$ there will be many rational approximations R_{nm}. We can now ask is it possible to find one rational approximation amongst this class which minimises the maximum error. It is possible to obtain results analogous to the Chebyshev equi-oscillation theorem and also de la Vallee Poussin but we refer the interested reader to Ralston (1965). Similarly, there is a procedure which corresponds to the Remes algorithm which determines the rational minimax approximation iteratively and is usually called Maehly's algorithm. This requires the solution of *non-linear* equations at each step. Unfortunately, due to the non-linearity of Maehly's algorithm there may not be a unique solution, but, it is possible to show that there is only one solution which does not introduce additional singularities into the required rational approximation. The determination of this unique rational approximation is very difficult and beyond the scope of this text.

Execise 3.7.4 (Computational) Use the program **APPROX** to contruct a rational approximation for $\exp(x)$, $|x| \leq 1$, by interpolation at five equally spaced points. Examine the error between the computed approximation and the function and adjust the interpolation points to try to reduce the maximum error. Recompute a new rational approximation based on the new interpolation points and examine the error function which is produced.

> 3.7.2 Padé approximation

Just as it is possible to consider a truncated Taylor series of a continuous function as an approximation, the equivalent expression in terms of a rational expansion is called a Padé approximation. To illustrate this consider the following example.

Example 3.7.6
We shall construct a Padé approximation for $\exp(x)$ which is of the form $R_{1,1}$, i.e.

$$\exp(x) \approx \frac{a_0 + a_1 x}{b_0 + x},$$

where the coefficients a_0, a_1 and b_0 are selected so that the function and the approximation have as many derivatives as possible equal when $x = 0$. Firstly, we re-arrange this expression to give

$$(b_0 + x)\exp(x) = a_0 + a_1 x \tag{3.7.2}$$

then when $x = 0$, i.e.

$$b_0 \exp(0) = b_0 = a_0. \tag{3.7.3}$$

Next we differentiate equation (3.7.2) to obtain

$$(b_0 + x)\exp(x) + \exp(x) = a_1$$

and then set $x = 0$ to produce

$$b_0 + 1 = a_1. \tag{3.7.4}$$

As we have three parameters at our disposal we can repeat this procedure once more to give

$$(b_0 + x)\exp(x) + \exp(x) + \exp(x) = 0$$

at $x = 0$ which gives $b_0 = -2$. Substituting back into equation (3.7.4) gives $a_1 = -1$ and into equation (3.7.3) to get $a_0 = -2$. Therefore, we have the Padé approximation

$$\exp(x) \approx \frac{-2 - x}{-2 + x}.$$

This approximation can be compared with the rational approximation obtained by interpolation at $x = -1$, $x = 0$ and $x = 1$ in example 3.7.2 which was

$$R_{1,1}(x) = \frac{-2.1639534 - x}{-2.1639534 + x}.$$

The errors associated with these two rational approximations is shown in figure 3.7.4.

Exercise 3.7.5 (Analytical and Computational) Compare the rational approximation $R_{1,3}$, $R_{2,2}$ and $R_{3,1}$ determined by interpolation at equally spaced points in $(-1, 1)$ with the corresponding Padé approximation.

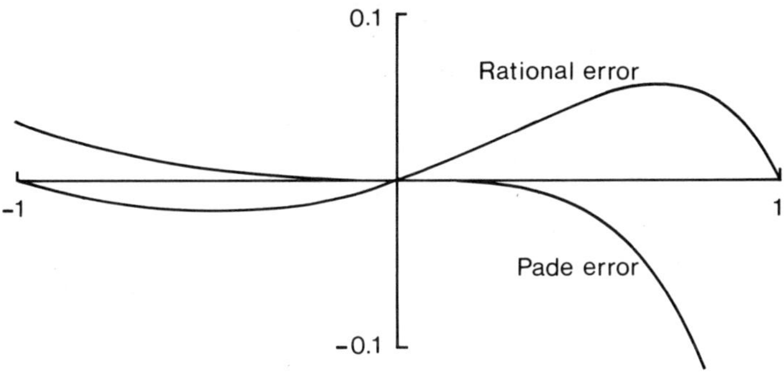

Figure 3.7.4 Padé and rational, $R_{1,1}$, errors for e^x.

> 3.8 Applications and conclusions

We now return to the problem posed at the beginning of this chapter; how does a computer evaluate sin(1.4) or exp(−3.456)? Different manufacturers have different methods but we shall illustrate the basic principles which are used. In general no approximation will be valid for all values of the argument and it is usual to define a principal range and then reduce the evaluation for a particular argument to this range. For example, we could use the periodicity of the trigonometric function to produce a principal range of $(0, 2\pi)$, in fact we can use other smaller principal ranges as we now demonstrate. One other point to notice is the heavy emphasis placed on binary arithmetic which is the basic form of computer arithmetic.

Example 3.8.1 (Analytical)
As a first example we shall illustrate how the function $\sin(x)$, x in radians, is evaluated. (The actual implementation will differ from machine to machine but the basic idea is common to all machines.)

1. If $|x| \leq 10^{-8}$ then return $\sin(x) \approx x$.

2. Reduce the argument by periodicity so that $|x| \leq \frac{1}{2}\pi$ and set $t = 2x/\pi$; then

$$\sin(x) \quad = \quad t \sum_{n=0}^{n=\infty} A_n T_n^*(t^2),$$

$$\cos(x) = \sum_{n=0}^{n=\infty} B_n T_n^*(t^2),$$

where T_n^* is the nth modified Chebyshev polynomial and the coefficients A_n and B_n are given in table 3.8.1.

Table 3.8.1 Modified Chebyshev coefficients for $\sin(x)$ and $\cos(x)$.

n	A_n	B_n
0	1.276278962	0.472001216
1	−0.285261569	−0.499403258
2	0.009118016	0.027992080
3	−0.000136587	−0.000596695
4	0.000001185	0.000006704
5	−0.000000007	−0.000000047

Accuracy. For $|x| \le \frac{1}{2}\pi$ the absolute error in the evaluation of $\sin(x)$ does not exceed 0.25×10^{-8} and for $\cos(x)$ it does not exceed 0.42×10^{-7}.

Example 3.8.2 (Analytical)
Evaluate $\sin(1.4)$. The required argument lies within $(-\frac{1}{2}\pi, \frac{1}{2}\pi)$ therefore we set $t = (2 \times 1.4)/\pi \approx 0.891267681$. Next we use the recurrence relations

$$\begin{aligned} T_0^*(t^2) &= 1, \qquad T_1^*(t^2) = 2t^2 - 1, \\ T_{n+1}^*(t^2) &= 2(2t^2 - 1)T_n^*(t^2) - T_{n-1}^*(t^2). \end{aligned}$$

Then

$$\begin{array}{ll} T_0^*(t^2) = 1, & T_1^*(t^2) = 0.588716160, \\ T_2^*(t^2) = -0.306826567, & T_3^*(t^2) = -0.949983676, \\ T_4^*(t^2) = -0.811714915, & T_5^*(t^2) = -0.005755699. \end{array}$$

from which we obtain $\sin(1.4) \approx 0.985449729$ which has absolute error 0.78×10^{-9}. Similarly, $\cos(1.4) \approx 0.169967143$ which has absolute error 0.36×10^{-10}.

Exercise 3.8.1 (Calculator) Evaluate $\sin(1.0)$ using the algorithm described in example 3.8.1.

Example 3.8.3 (Analytical)
The evaluation of $\exp(x)$.

1. If $x < -180$ return a value 0.
2. Divide x by $\log_e(2)$ and write
$$x = \left(4a - b - \tfrac{1}{16}c\right)\log_e(2) - d,$$
where a, b and c are integers with $0 \le b \le 3$, $0 \le c \le 15$ and $0 \le d < \frac{1}{16}\log_e(2)$, then
$$\exp(x) = 16^a 2^{-b} 2^{-c/16}\exp(-d).$$
3. Compute $\exp(-d)$ using a minimax approximation polynomial approximation of degree 6.
4. Multiply $\exp(-d)$ by $2^{-c/16}$ then halve the result b times.
5. Multiply by 16^a by adding the hexadecimal exponent a to the exponent of the previous operation to give $\exp(x)$.

Accuracy. If $-1 < x < 1$ then the maximum relative error in $\exp(x)$ is less than 0.21×10^{-15}, otherwise it is less than 0.43×10^{-15}.

Example 3.8.4 (Analytical)
The natural logarithm is evaluated using a rational approximation.

1. If $x = 1$ return $\log_e(x) = 0$.
2. Determine an integer n such that $2^{n-1} \le x \le 2^n$ and then put $r = 2^{-n}x$ $(\frac{1}{2} \le r < 1)$. Set $u = (r - \frac{1}{2}\sqrt{2})/(r + \frac{1}{2}\sqrt{2})$ and find
$$\begin{aligned} v &= \log_e\left(\frac{1+u}{1-u}\right) \\ &\approx u\left(\frac{20790 - 21545.27u^2 + 4223.9187u^4}{10395 - 14237.635u^2 + 4778.8377u^4 - 230.41913u^6}\right) \end{aligned}$$
then
$$\log_e(x) \approx \left(n - \tfrac{1}{2}\right)\log_e 2 + v.$$

Accuracy. The maximum absolute error in $\log_e(x)$ is less than 1.02×10^{-10}.

Exercise 3.8.2 (Calculator/Computational) Use the algorithms given in examples 3.8.3 and 3.8.4 to evaluate exp(3.456) and $\log_e(3.456)$.

Not all standard functions use polynomial approximations. For example, the evaluation of the square root function uses an iterative approach.

Example 3.8.5 (Analytical)
The evaluation of $\sqrt{x}$.

1. If $x = 0$ then return 0.

2. Let $x = 16^{2p-q} f$, where p is an integer, $q = 0$ or 1 and $\frac{1}{16} < f < 1$ then

$$\sqrt{x} = 16^p 4^{-q} \sqrt{f}.$$

3. Take as a first approximation for $\sqrt{f}$

$$y_0 = 16^p 4^{-q} \frac{(0.14605 + 1.8342f)}{(0.98026 + f)}.$$

4. Apply the Newton–Raphson iteration

$$y_{n+1} = \tfrac{1}{2}\left(y_n + \frac{x}{y_n}\right)$$

twice. The final iteration is taken as $y_2 = \frac{1}{2}(y_1 - \frac{x}{y_1}) + \frac{x}{y_1}$ in order to minimize computational errors.

Accuracy. The maximum relative error is 0.48×10^{-6}.

Example 3.8.6 (Calculator)
Let us determine $\sqrt{3}$. We can write $3 = 16^{2-1}\,0.1875$, therefore $p = 1$, $q = 1$ and $f = 0.1875$. Using these values we obtain $y_0 = 1.67829862$, $y_1 = 1.73291159$, $y_2 = 1.73205102$ and $y_3 = 1.73205081$. This last value is correct to eight decimal places.

Exercise 3.8.3 (Calculator) Use the algorithm outlined in example 3.8.5 to determine $\sqrt{1.76}$. Square the result and hence estimate the error. Compare the error with the bound which is given.

In this chapter we have seen how it is possible to construct both polynomial and rational approximations for a given function. It is difficult to say which type of approximation will in general be the best; this will depend on the accuracy required and the frequency with which the approximation is going to be used. It is arguable that in most cases a Chebyshev approximation will be sufficient for most purposes. The determination of a minimax approximation requires considerable effort, but once determined only the coefficients in the relevant polynomial need to be stored and evaluation is particularly easy.

> Chapter 4

> Least Squares Approximation

Of course the first thing to do was to make a grand survey of the country she was going to travel through. "It's something like learning geography", thought Alice, as she stood on tiptoe in hopes of being able to see a little farther.

Lewis Carroll.

> 4.1 Introduction

In Chapter 2 the problem of fitting a function through a set of data points was considered. It was shown that as the number of points increases then polynomial approximation becomes more oscillatory. In order to overcome this problem the concept of spline interpolation was introduced whereby lower order polynomials were employed. In this chapter we shall consider an alternative approach; to do this consider the following example.

Example 4.1.1 (Analytical)
Newton's law of cooling states that the rate at which a body loses heat is proportional to the difference between the temperature of the body and the surrounding ambient temperature T_0. Let T be the temperature at time t; then we can write

$$\frac{dT}{dt} = -K(T - T_0),$$

where K is some constant, which can be integrated to give

$$T(t) = T_0 + A\exp(-Kt). \tag{4.1.1}$$

In an experiment the ambient temperature was found to be 10 °C, and the data in table 4.1.1 were observed; how can we find the most suitable value for the constant K? The values of $T - 10$ are plotted against t in figure 4.1.1, as are the curves $T_0 + A\exp(-Kt)$ for an assortment of A and K. The problem of minimizing the error between the data points and any one member of this family of curves is quite difficult.

Table 4.1.1 Data obtained for example 4.1.1.

t (min)	T °C	$\log_e(T-10)$
1	65	4.007333
2	48	3.637586
3	30	2.995732

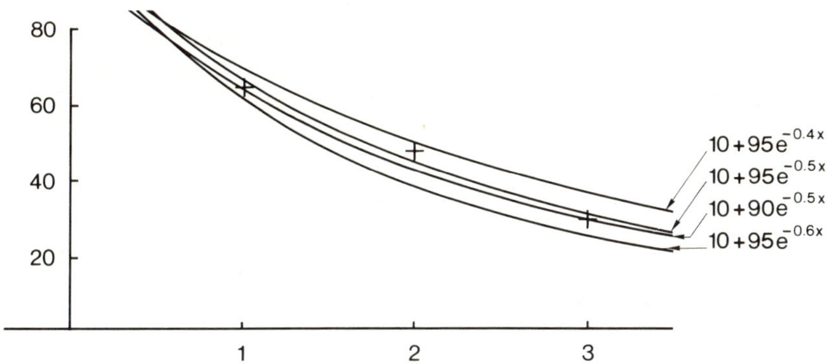

Figure 4.1.1 Values from table 4.1.1 and $10 + A\exp(-Kt)$.

As an alternative we can rewrite equation (4.1.1) as

$$\log_e(T - T_0) = B - Kt, \qquad B = \log_e A,$$

and plot the new variable $y = \log_e(T - T_0)$ against t. We then have a linear relationship between y and t, and the slope of this line is related to the value K we require. For a particular choice of B and K it is unlikely that the selected line will pass through all the data points and so we write the error at a point t_i as

$$e_i = y(t_i) - y_i, \tag{4.1.2}$$

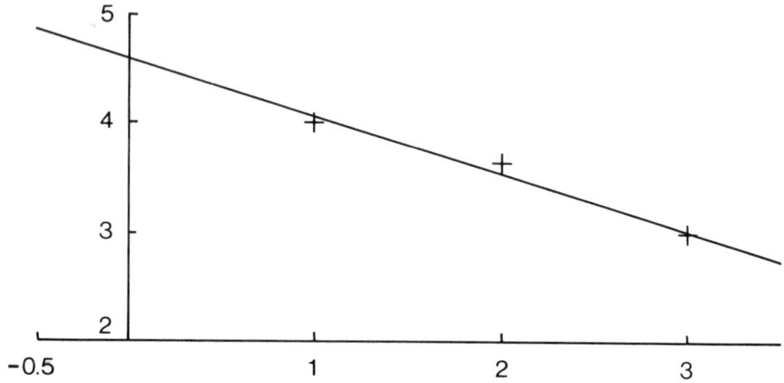

Figure 4.1.2 Plot of t against $\log_e(T-10)$ for table 4.1.1.

where $y(t_i)$ is the value determined from the particular choice of the parameters and y_i is the observed value at t_i. In order to obtain the 'best possible' fit we could attempt to select the parameters B and K to minimize the sum of squares, S, of the errors. In this case

$$S = (B-K-4.007333)^2+(B-2K-3.637586)^2+(B-3K-2.995732)^2. \tag{4.1.3}$$

This function has a turning point where $\partial S/\partial B = \partial S/\partial K = 0$ which gives two linear equations for K and B. Solving these equations produces the least squares approximation for the given data as

$$\log_e(T-T_0) = 4.558485 - 0.5058005t, \tag{4.1.4}$$

or equivalently

$$T(t) = 10 + 95.43878\exp(-0.5058005t). \tag{4.1.5}$$

The corresponding sum of squares, S, given by equation (4.1.3) is then 1.234×10^{-2}. Notice that (4.1.4) is a linear least squares fit for the function $\log_e(T-T_0)$ but that (4.1.5) is not necessarily a non-linear least squares fit for T. We shall discuss linear least squares in some detail; non-linear least squares is far more difficult. A detailed discussion is given by Stoer and Burlisch (1980).

> 4.2 Linear least squares

In the last section we used a simple demonstration which fitted a straight line by minimizing the sum of squares of the associated errors. We will now extend this approach to fitting a linear function to a general set of values and then generalize the method to fit a linear combination of functions.

Let us suppose that we have been given a set of tabular values (x_j, y_j), $j = 1, \ldots, m$, and that we wish to find a linear function,

$$y = c_0 + c_1 x, \tag{4.2.1}$$

which minimizes some overall measure of the difference between the tabular values y_j and the corresponding $y(x_j)$ as provided by the approximating function. There is no unique way of providing a measure of the discrepancy between y_j and $y(x_j)$. For example, we could use the sum of the magnitudes of the errors to compare different values of c_0 and c_1, i.e.

$$\sum_{j=0}^{j=m} |y(x_j) - y_j| = \sum_{j=0}^{j=m} |c_0 + c_1 x_j - y_j|. \tag{4.2.2}$$

It is possible to determine values of c_0 and c_1 which minimize this functional but it requires linear programming techniques beyond the scope of this text. (If such an approximation is found it is called an l_1 approximation.) However, in practice it is easier to minimize the sum of squares of the errors, i.e.

$$S(c_0, c_1) = \sum_{j=0}^{j=m} (c_0 + c_1 x_j - y_j)^2. \tag{4.2.3}$$

This is called the least squares fit or an l_2 approximation. We could select values for c_0 and c_1 at random and compute the value of $S(c_0, c_1)$ but a much simpler way is to use the calculus.

Exercise 4.2.1 (Computational) Load the program **LSQ** and change the prompt to **Select data**. Now change the current option to **Data: from keys** and clear the existing data. Select the prompt **Data input format: tabular** and input the entries in the first and third columns of table 4.1.1. (Recall that $\log_e(65 - 10)$ can be input as LN(55).) The data points can be plotted on suitable axes by using the option **Data: plot&/or edit** which should produce a screen display like figure 4.1.2. (Recall that these data can be stored in a file using the option **Data: store on file** but you will need a disc of your own which is not write protected.) Exit from this set of options by using the prompt **Data: quit** and when the option shows **Method: Linear** press RETURN. Select the prompt **Option: interactive fitting** and a default line will appear on the screen together with the current sum of squares of deviations (SSD) as given by equation (4.2.3). The position and slope of the current line can be changed by using the cursor keys.

1. ↑ ↓ shift the current line with respect to the axis $y = 0$.

2. ← → rotate the current line about the middle of the screen.

3. <> increase or decrease the increment by which the line is changed.

As the line is moved the sum of squares of deviations (SSD) is changed and displayed in the top window. It is also possible to examine the current values of the coefficients in equation (4.2.3), c_0 and c_1, by selecting the prompt **Option: edit/inspect coeff**. Try to adjust the line to minimize the sum of squares of deviations.

Exercise 4.2.1 illustrates some of the problems in finding a least squares approximation by trial and error; therefore, we will now consider a more direct approach. The functional $S(c_0, c_1)$, given by equation (4.2.3), has turning points whenever

$$\frac{\partial S}{\partial c_1} = \frac{\partial S}{\partial c_0} = 0, \tag{4.2.4}$$

therefore, necessary conditions for a minimum are

$$\begin{aligned} \sum_{j=0}^{j=m} x_j(c_0 + c_1 x_j - y_j) &= 0, \\ \sum_{j=0}^{j=m} (c_0 + c_1 x_j - y_j) &= 0. \end{aligned}$$

Finally, these equations can be rewritten in the form

$$c_0 \sum_{j=0}^{j=m} x_j + c_1 \sum_{j=0}^{j=m} x_j^2 = \sum_{j=0}^{j=m} x_j y_j, \tag{4.2.5}$$

$$(m+1)\, c_0 + c_1 \sum_{j=0}^{j=m} x_j \;=\; \sum_{j=0}^{j=m} y_j .$$

For the example in section 4.1 these equations reduce to the linear equations

$$\begin{pmatrix} 6 & 14 \\ 3 & 6 \end{pmatrix} \begin{pmatrix} c_0 \\ c_1 \end{pmatrix} = \begin{pmatrix} 20.269701 \\ 10.640651 \end{pmatrix}.$$

The solution of these equations is $c_0 = 4.558485$ and $c_1 = -0.5058005$ as in section 4.1. These coefficients produce the fitted linear function $y = 4.558485 - 0.5058005t$.

Exercise 4.2.2 (Computational) Load the program **LSQ** and select the prompts **Method: Linear** and **Option: calculate coeffs**. The program will now determine the least squares linear fit by using equation (4.2.6) and then return to this prompt. It is possible to display the fitted function by selecting the prompt **Option: plot** and examine the computed coefficients using the prompt **Option: edit/inspect coeffs**. Compare the computed least squares fit with the results of exercise 4.2.1.

Exercise 4.2.3 (Computational) Repeat exercise 4.2.2 but investigate the sensitivity of the computed line with respect to small preturbations of the data points. Is this problem well conditioned?

We could generalize the above argument to consider functions of the form

$$y(x) = c_0 + c_1 x + c_2 x^2 + \ldots + c_n x^n$$

but these can in turn be considered as a special case of

$$y(x) = c_0\phi_0 + c_1\phi_1 + \ldots + c_n\phi_n = \sum_{i=0}^{i=n} c_i \phi_i(x), \qquad (4.2.6)$$

where $\phi_i(x)$, $i = 0, \ldots, n$, are any suitable basis functions. It is clear that $n \leq m$ but if $n = m$ the problem reduces to constructing an interpolating polynomial through the data points. As before we attempt to determine coefficients c_i, $i = 0, \ldots, n$, in order to minimize the functional

$$S(c_0, c_1, \ldots, c_n) = \sum_{j=0}^{j=m} (y_j - y(x_j))^2. \qquad (4.2.7)$$

At a turning point we shall have

$$\frac{\partial S}{\partial c_i} = 0, \qquad i = 0, \ldots, n, \qquad (4.2.8)$$

which produces a system of $n+1$ linear equations in the unknown coefficients c_i, $i = 0, \ldots, n$. The equations (4.2.7) may be rewritten as

$$S(c_0, \ldots, c_n) = \sum_{j=0}^{j=m} \left(y_j - \sum_{i=0}^{i=n} c_i \phi_i(x_j) \right)^2 . \qquad (4.2.9)$$

Therefore, differentiating with respect to c_k produces

$$\sum_{j=0}^{j=m} \phi_k(x_j) \left(y_j - \sum_{i=0}^{i=n} c_i \phi_i(x_j) \right) = 0, \quad k = 0, \ldots, n. \qquad (4.2.10)$$

which can in turn be re-arranged to give

$$\sum_{i=0}^{i=n} \left(\sum_{j=0}^{j=m} \phi_k(x_j) \phi_i(x_j) \right) c_i = \sum_{j=0}^{j=m} \phi_k(x_j) y_j, \quad k = 0, \ldots, n. \quad (4.2.11)$$

Equivalently, we may rewrite these equations as

$$\mathbf{Ac} = \mathbf{b},$$

where the components of the matrix $\mathbf{A}$ are given by

$$a_{ki} = \sum_{j=0}^{j=m} \phi_k(x_j) \phi_i(x_j)$$

and those of the vector $\mathbf{b}$ by $b_k = \sum_{j=0}^{j=m} \phi_k(x_j) y_j$. Notice that $\mathbf{A}$ is independent of the values y_j, $j = 0, \ldots, m$. The system of linear equations (4.2.11) are called *normal equations* and can be solved by Gaussian elimination without the need to partially pivot since the matrix of coefficients is symmetric and positive definite.

Exercise 4.2.4 (Computational) Load the program **LSQ** for which the default data points are given in table 2.1.1. When the prompt shows **Select method** press RETURN and then use the option **Method: Linear** to determine a linear least squares approximation for these data points using the option **interactive fitting** and then plot it. Determine the least squares fit by using the option **calculate coeffs** and compare the computed results. Now select the option **Method: Polynomial** and fit polynomials of degree 2, 3 and 4. (Why is it not possible to fit higher order polynomials to these data points?) Modify the data points using the option **edit/inspect points** and observe the way the fitted function changes.

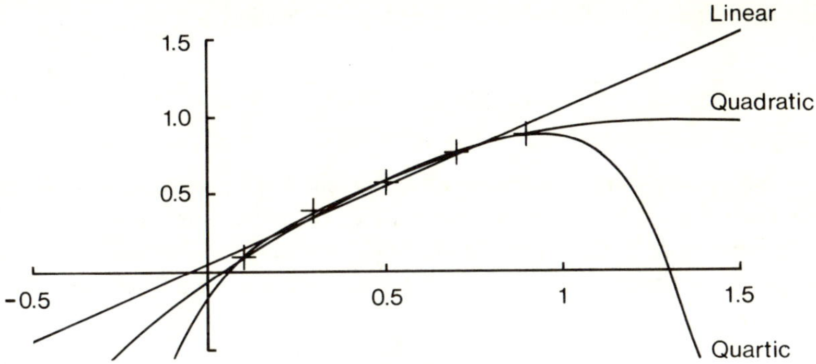

Figure 4.2.1 Linear and polynomial fitting for table 2.1.1.

Exercise 4.2.5 (Computational) Repeat exercise 4.2.4 but fitting
(i) Chebyshev polynomials of degree 2, 3 and 4.
(ii) Fourier series of order 1 and 2.
(iii) The functions 1, $\exp(x)$, $\exp(2x)$, ...
For (iii) what is the maximum number of such terms which can be used?

Exercise 4.2.6 (Computational) Repeat exercise 4.2.4 but try specifying any suitable set of basis functions of your choice. (Recall that they must be linearly independent.)

Exercise 4.2.7 (Analytical) Show that the functional, S, given in equation (4.2.7) may be written in the matrix form

$$S(c_0, \ldots, c_n) = (\mathbf{Nc} - \mathbf{y})^T(\mathbf{Nc} - \mathbf{y}),$$

where $\mathbf{N}$ is a rectangular matrix for which $N_{ij} = \phi_i(x_j)$, $\mathbf{y}$ and $\mathbf{c}$ are vectors whose components are y_j, $j = 0, 1, \ldots, m$ and c_i, $i = 0, \ldots, n$, respectively. By differentiating this form directly show that the normal equations become

$$\mathbf{N}^T\mathbf{Nc} = \mathbf{N}^T\mathbf{y},$$

and deduce that these equations are identical to the equations (4.2.11). Furthermore, show that the matrix $\mathbf{N}$ is symmetric and positive definite.

The matrix $\mathbf{A}$ in the normal equations (4.2.11) will in general be full but if the basis functions $\phi_i(x)$, $i = 0, \ldots, n$, satisfy

$$\sum_{j=0}^{j=m} \phi_r(x_j)\phi_s(x_j) = 0, \qquad \text{if } r \neq s \tag{4.2.12}$$

then all the off-diagonal entries of **A** vanish. If this is the case then the functions $\phi_i(x)$, $i = 0, \ldots, n$, are said to be *orthogonal with respect to the points* x_j, $j = 0, \ldots, m$. Under this condition the normal equations reduce to a diagonal system and the coefficients of (4.2.6) are given by

$$c_i = \frac{\sum_{j=0}^{j=m} \phi_i(x_j)y_j}{\sum_{j=0}^{j=m} \phi_i(x_j)^2}, \; i = 0, \ldots, n. \tag{4.2.13}$$

Given any set of linearly independent functions it is possible to generate another set which satisfies (4.2.12) but we shall consider an alternative approach in a later section.

Exercise 4.2.8 (Computational) Load the program **LSQ** and input the data file **SPLINE**. Compare linear, quadratic and cubic least squares fits for these data. Explain why the fitted function is so poor. Repeat this exercise for the datafile **SIN**.

> 4.3 Continuous least squares approximation

In the last section we discussed the problem of fitting a function of the form

$$y(x) = c_0\phi_0(x) + \ldots + c_n\phi_n(x), \tag{4.3.1}$$

to a set of discrete points. This idea can be generalised to consider the approximation of a given function $f(x)$, $|x| \leq 1$, by expressions of this form. In order to do this we consider the functional

$$S(c_0, \ldots, c_n) = \int_{-1}^{1} \left(f(x) - \sum_{i=0}^{i=n} c_i\phi_i(x) \right)^2 dx \tag{4.3.2}$$

as a measure of the error between the function and its approximation. It is also possible to consider a weighted measure of the discrepancy by defining

$$S(c_0, \ldots, c_n) = \int_{-1}^{1} w(x) \left(f(x) - \sum_{i=0}^{i=n} c_i\phi_i(x) \right)^2 dx, \tag{4.3.3}$$

where $w(x)$ is a suitable weight function. We have assumed, without loss of generality, that the function $f(x)$ is defined on the interval $[-1,1]$; if not then we can always transform it to be so. As before we differentiate $S(c_0,\ldots,c_n)$ with respect to c_k to produce the corresponding normal equations, i.e.

$$\sum_{i=0}^{i=n}\left(\int_{-1}^{1} w(x)\phi_k(x)\phi_i(x)\,dx\right)c_i = \int_{-1}^{1} w(x)\phi_k(x)f(x)\,dx,\ k=0,\ldots,n. \tag{4.3.4}$$

This produces a system of linear equations in the form $\mathbf{Ac}=\mathbf{b}$, where

$$a_{ki} = \int_{-1}^{1} w(x)\phi_k(x)\phi_i(x)\,dx$$

and

$$b_k = \int_{-1}^{1} w(x)\phi_k(x)f(x)\,dx.$$

However, if the basis function $\phi_i(x)$, $i=0,\ldots,n$, are orthogonal with respect to the weight function $w(x)$, i.e.

$$\int_{-1}^{1} w(x)\phi_i(x)\phi_j(x)\,dx = 0, \quad i\neq j,$$

then the normal equations are particularly simple and the coefficients c_i, $i=0,\ldots,n$, are given by

$$c_i = \frac{\int_{-1}^{1} w(x)\phi_i(x)f(x)\,dx}{\int_{-1}^{1} w(x)\phi_i(x)^2\,dx}. \tag{4.3.5}$$

Example 4.3.1 (Analytical)

The Chebyshev polynomials form an orthogonal basis. (See section 3.4 and example 3.5.1.) Hence, we can construct the Chebyshev least squares approximation for a function $f(x)$, $x\in[-1,1]$ as follows. We consider an approximation in the first $(n+1)$ Chebyshev polynomials given by

$$f(x)\approx\sum_{i=0}^{i=n} c_iT_i(x).$$

Then from equation (4.3.5) the least squares coefficients are given by

$$c_0 = \frac{1}{\pi}\int_{-1}^{1}\frac{f(x)}{\sqrt{(1-x^2)}}\,dx, \quad c_i = \frac{2}{\pi}\int_{-1}^{1}\frac{T_i(x)f(x)}{\sqrt{(1-x^2)}}\,dx.$$

Exercise 4.3.1 (Analytical) Show that for the coefficients given in example 4.3.1 the functional $S(c_0, c_1, \ldots, c_n)$ given in equation (4.3.3), has minimum value

$$\int_{-1}^{1} \frac{f^2(x)}{\sqrt{(1-x^2)}}\, dx - \tfrac{1}{2}\pi(2c_0^2 + c_1^2 + \ldots + c_n^2).$$

Exercise 4.3.2 (Analytical) Show that the Fourier series approximation

$$f(x) \approx \frac{a_0}{2} + \sum_{k=1}^{k=n}(a_k \cos kx + b_k \sin kx)$$

with the Fourier coefficients defined by

$$a_k = \frac{1}{\pi}\int_{-\pi}^{\pi} f(x)\cos kx\, dx, \qquad b_k = \frac{1}{\pi}\int_{-\pi}^{\pi} f(x)\sin kx\, dx,$$

is a least squares approximation for $f(x)$, with unit weight, in $|x| \le \pi$. (The question of the convergence of the Fourier series as $n \to \infty$ will not be discussed here. For a full discussion see Churchill (1963).)

> 4.4 Discrete least squares approximation

In the last section we saw how it was possible to construct a continuous least squares approximation for a given function, however, to do so it was necessary to evaluate the coefficients by performing numerous integrations. We will now return to consider how it is possible to construct a discrete variant which avoids this problem.

In section 4.2 we saw that given a set of points (x_j, y_j), $j = 0, \ldots, m$, we could construct a linear least squares fit, $\sum_{i=0}^{i=n} c_i\phi_i(x)$, by solving the normal equations given by

$$\sum_{i=0}^{i=n}\left(\sum_{j=0}^{j=m}\phi_k(x_j)\phi_i(x_j)\right)c_i = \sum_{j=0}^{j=m}\phi_k(x_j)y_j,\ k = 0, \ldots, n, \qquad (4.4.1)$$

but if the basis functions are orthogonal with respect to the tabular points, i.e.

$$\sum_{j=0}^{j=n}\phi_k(x_j)\phi_i(x_j) = 0, \qquad i \ne k, \qquad (4.4.2)$$

then the coefficients in the least squares approximation are given by the simple form

$$c_i = \frac{\sum_{j=0}^{j=m} \phi_i(x_j) y_j}{\sum_{j=0}^{j=m} \phi_i(x_j)^2}, \qquad i = 0, \ldots, n. \tag{4.4.3}$$

Therefore, given any function $f(x)$ we can construct the least squares approximation to it if $(m+1)$ points, $(x_j, f(x_j))$, $j = 0, \ldots, m$, are known such that equation (4.4.2) is satisfied. Conversely, if we have $m+1$ points which satisfy this condition then the interpolating polynomial constructed in this way will be the least squares approximation.

In order to illustrate this approach consider the Chebyshev approximation given by

$$f(x) \approx {\sum_{i=0}^{i=n}}' a_i T_i(x), \quad -1 \leq x \leq 1, \tag{4.4.4}$$

where the $'$ denotes the fact that the first coefficient is halved. Then if the tabulation points x_j, $j = 0, \ldots, m$, are selected to be the zeros of T_{m+1}, i.e.

$$x_j = \cos\left(\frac{(2j+1)}{(m+1)} \frac{\pi}{m}\right), \; j = 0, \ldots, m, \tag{4.4.5}$$

it is possible to show[1] that

$${\sum_{j=0}^{j=m}}'' T_r(x_j) T_s(x_j) = \begin{cases} 0, & r \neq s, \\ \frac{1}{2}m, & r = s \neq 0, \\ m, & r = s = 0, \end{cases} \tag{4.4.6}$$

where the $''$ denotes that both the first and last terms in the summation, for any r or s, are to be halved. In this case the coefficients in equation (4.4.4) are given by

$$a_i = \frac{2}{m} {\sum_{j=0}^{j=m}}'' f(x_j) T_i(x_j). \tag{4.4.7}$$

This is the discrete Chebyshev least squares approximation. If $n = m$, i.e. the number of tabulation points is equal to the number of coefficients, then the $'$ in equation (4.4.4) is replaced by $''$ for an exact fit. Notice that the tabulation points x_j are not equally spaced.

[1] See Cheney (1966).

Example 4.4.1 (Analytical)
Let us consider the function $\exp(x)$, $x \in [-1,1]$. The Chebyshev series which represents this function is given by

$$\begin{aligned}\exp(x) = 1.266066T_0 &+ 1.130318T_1 + 0.271495T_2 + 0.044337T_3 \\ &+ 0.005474T_4 + 0.000543T_5 + \dots .\end{aligned}$$

The Chebyshev discrete least squares approximation of degree 2 which is constructed by using least squares at the zeros of T_4 is given by

$$1.266066T_0 + 1.130315T_1 + 0.271450T_2.$$

The maximum difference between these functions is 5.045×10^{-2} and the sum of squares of deviations is 3.835×10^{-3}.

We can compare the coefficients of a Chebyshev series and those of the discrete Chebyshev least squares as follows. Recall that the coeff-cients of the Chebyshev series, c_i, are given by

$$c_i = \frac{2}{\pi}\int_0^{\pi} f(\cos\theta)\ \cos i\theta\ d\theta$$

and if we apply the composite trapezium rule to this integral with $m+1$ points we find

$$\begin{aligned}c_i &\approx \frac{2}{m}\sum_{k=0}^{k=m}{}'' f(\cos\theta_k)\ \cos i\theta_k, \qquad \theta_k = \frac{k\pi}{m}, \\ &= \frac{2}{m}\sum_{k=0}^{k=m}{}'' f(x_k)T_i(x_k) = a_i,\end{aligned}$$

from (4.4.7), i.e. the Chebyshev series of $f(x)$ is the Fourier series of $f(\cos(x))$.

Exercise 4.4.1 (Computational) Load the program **LSQ** and change the prompt to **Select function**. When the current option shows **Function: disabled** change it to **enabled** and press RETURN. Now change the option to **Function: inspect/edit** and change the function to e^x, $|x| \le 1$. Quit this section of the program and change the prompt to **Select data**. Use the option **Data: standard grid** to select a uniform grid of four points and then quit. The program should now return to the prompt **Method: Linear**; change the option to **Polynomial** and fit a polynomial of degree 2. Compare the result with a Chebyshev least squares fit of the same degree. What do you deduce? Increase the number of points and repeat the comparison. Return to the option **Select data** and use the zeros of $T_4(x)$ as data points, construct a least squares fit for e^x, $|x| \leq 1$, based on these points and then compare it with your results for equally spaced points.

Exercise 4.4.2 (Computational) Construct an interpolating polynomial approximation for e^x, $|x| \leq 1$, which passes through six uniformly spaced points. (Use **INTERP**). Compare the result with least squares polynomial fits for polynomials of degree 3, 4 and 5. Repeat this exercise using the zeros of $T_6(x)$. What do you conclude?

Exercise 4.4.3 (Computational) Construct the minimax cubic polynomial approximation for e^x, $|x| \leq 1$, using the Remes algorithm in **APPROX**. Compare it with least squares Chebyshev approximations of order 3,4 and 5. Estimate the relative efficiency of determining these approximations.

> Chapter 5

> Integration

'Quid pro Quadrature?'

Although many functions are known to be integrable, that is to say they have an indefinite integrand, there may be no expression for the integrand in terms of elementary functions. Perhaps the simplest example of this is the integral

$$\int_0^1 \exp(-x^2)\, dx.$$

In this chapter we will see how it is possible to determine an estimate for such integrals. To do this we first find an approximation for the integrand, integrate the approximation analytically in order to estimate the integral and then estimate the overall error. This process is called *quadrature*.

> 5.1 Simple quadrature

Let $f(x)$ be a continuous function; then we may use its Taylor series expansion about $x = a$ to approximate it, i.e.

$$f(x) = f(a) + (x-a)f'(a) + \frac{(x-a)^2}{2} f''(a) + \ldots + \frac{(x-a)^n}{n!} f^{(n)}(a) + \ldots. \tag{5.1.1}$$

Therefore,

$$\int_a^b f(x)\, dx \quad = \quad f(a)\,[x-a]_a^b + f'(a)\left[\frac{(x-a)^2}{2}\right]_a^b + \ldots$$

$$= \quad (b-a)f(a) + \tfrac{1}{2}(b-a)^2 f'(a) + \ldots .$$

By adding in sufficient terms, and providing $(b-a)$ is sufficiently small, we might hope to estimate the integral. If we were to replace the expansion (5.1.1) by a Taylor series with remainder we might even be able to estimate the error but as we have seen the Taylor series expansion is not the best approximation available. As an alternative, for any set of points x_j, $j = 0, \ldots, m$, we can find the Lagrangian polynomial approximation

$$p_n(x) = \sum_{j=0}^{j=n} L_j(x) f(x_j), \tag{5.1.2}$$

which has error

$$e_n(x) = f(x) - p_n(x) = \frac{f^{(n+1)}(\theta)}{(n+1)!} \prod_{j=0}^{j=n} (x - x_j), \tag{5.1.3}$$

where θ depends on x and lies between x_0 and x_n. Therefore,

$$\int_a^b f(x)\,dx = \int_a^b (p_n(x) + e_n(x))dx \tag{5.1.4}$$

$$= \int_a^b \sum_{j=0}^{j=n} L_j(x) f(x_j)\,dx + \int_a^b e_n(x)\,dx \tag{5.1.5}$$

$$= \sum_{j=0}^{j=n} \alpha_j f(x_j) + E_n, \tag{5.1.6}$$

where

$$\alpha_j = \int_a^b L_j(x)\,dx, \tag{5.1.7}$$

$$E_n = \frac{1}{(n+1)!} \int_a^b \prod_{j=0}^{j=n} (x - x_j)\, f^{(n+1)}(\theta)\,dx. \tag{5.1.8}$$

The approach now is to select the points x_j in order to minimize the error E_n. Notice that the quadrature formula (5.1.6) involves $(2n+2)$ parameters; the interpolation points x_j and the coefficients α_j, $j = 0, \ldots, n$, but that these values are not independent. For any arbitrary choice of interpolation points, usually called the abscissa, the coefficients, which are called the weights, are determined from equation (5.1.7). As a result we should be able to make the quadrature formula exact for polynomials of degree n since $n+1$ points determine a polynomial of degree n exactly. This is confirmed by the fact that e_n is proportional to $f^{(n+1)}$ which vanishes for all polynomials of degree less than $n+1$. Quadrature formulae of this type are called Newton–Cotes methods and will be considered in section 5.2. Alternatvely, we can generalize this approach and begin with the formula

$$\int_a^b f(x)\,dx = \sum_{j=0}^{j=n} \alpha_j f(x_j) + E_n \tag{5.1.9}$$

and then select values for all the x_j's and α_j's to make this a method of order higher than $n+1$. Such formulae are called Gaussian quadrature methods and they will be considered in section 5.3.

Example 5.1.1 (Computational)
The program **QUAD** is intended as a demonstration program for simple quadrature formulae, whereas **INTEGR** is intended for practical numerical quadrature. In order to illustrate how the choice and number of interpolation points influences the accuracy of numerical quadrature load the program **QUAD** for which the default problem is

$$\int_0^1 x^5\,dx.$$

Accept the prompts **Function: as set** and **Select data** by pressing RETURN. The prompt will now change to **Option: integrate**. Notice that in the graphical window three equally spaced points are placed on the default integrand x^5. This is confirmed in the bottom window where $N = 3$ is displayed. Furthermore, as three points are available it is possible to construct an interpolating polynomial of degree 2 which is confirmed by $n = 2$ in the top window. Now press RETURN. The bottom screen will display the estimate for the integral which has been determined as the **Current Estimate** together with an estimate of the quadrature error as determined by approximating equation (5.1.8). It is possible to examine the interpolating polynomial which has been used to approximate the integrand by changing the prompt to **Option: plot**, pressing RETURN and then accepting the option **Plot: approximation**. In addition, the difference between the integrand and the approximating polynomial can be highlighted using the option **Plot: shade** and the error can be displayed using the option **Plot: error**.

Exercise 5.1.1 (Computational) Load the program **QUAD** with the default problem

$$\int_0^1 x^5\,dx$$

and by changing the prompt **Select quadrature** to **Select data** investigate how the number of points in a uniform mesh influences the accuracy of the computed estimate for the integral. Why are six points sufficient to give an exact answer?

Example 5.1.2 (Computational)
Load the program **QUAD** to compute an approximation for

$$\int_{-1}^1 \exp(-x^2)\,dx$$

using a uniform grid of five points[1]. (Hint: Use the option **Select function** to change the integrand.) Return to the option **Select data** and then select **Data: plot &/or edit** and change the interpolation points to be at $x = 0, \pm 0.5, \pm 1$. Quit the data editor and select the prompt **Option: integrate**. An estimate for the value of the integral will now be determined based on these points together with an estimate of the quadrature error. Notice that the previous estimate remains visible in order that you can compare it.

Exercise 5.1.2 (Computational) Repeat example 5.1.2 for other sets of five interpolation points and try to determine the best possible choice, i.e. minimize the computed error estimate.

> 5.2 Newton–Cotes methods

Let us now return to consider Newton–Cotes formulae which are based on replacing the integrand by an interpolating polynomial. We begin by considering the simplest possible case where we replace the integrand, $f(x)$, by a linear approximation $p_1(x)$, $a \le x \le b$, i.e.

$$f(x) \approx p_1(x) = \frac{b-x}{b-a}f(a) + \frac{x-a}{b-a}f(b), \qquad (5.2.1)$$

[1]Some computers treat the expression $\mathrm{EXP}(-X^2)$ as $\mathrm{EXP}((-X)^2)$ whereas we require $\mathrm{EXP}(-(X)^2)$.

which is the Lagrange polynomial through $(a, f(a))$ and $(b, f(b))$. Therefore,

$$\begin{aligned}\int_a^b f(x)\,dx &= \int_a^b p_1(x)\,dx + E_1\\ &= \frac{f(a)}{(b-a)}\int_a^b (b-x)\,dx + \frac{f(b)}{(b-a)}\int_a^b (x-a)\,dx + E_1\\ &= \tfrac{1}{2}(b-a)[f(a)+f(b)] + E_1.\end{aligned}$$

This is the simple trapezium rule. We may use equation (5.1.3) to determine the error, E_1, in this approximation as

$$\int_a^b e_1(x)\,dx = \int_a^b \frac{f''(\theta(x))}{2!}(x-a)(x-b)\,dx = \frac{f''(\theta)}{2!}\int_a^b (x-a)(x-b)\,dx \tag{5.2.2}$$

where the mean value theorem for integration has been used to simplify the integral. Integrating the last term gives

$$E_1 = -\frac{(b-a)^3}{12} f''(\theta), \tag{5.2.3}$$

where $a \le \theta \le b$. The accuracy of the trapezium rule is proportional to the cube of the interval width, provided the integrand has a bounded second derivative, and if we require higher accuracy then it is necessary to include further interpolation points.

Exercise 5.2.1 (Analytical) Show that by sub-dividing the interval (a, b) into two sub-intervals such that $x_0 = a$, $x_1 = \frac{1}{2}(a+b)$, $x_2 = b$ and then replacing the integrand by a Lagrangian polynomial approximation of degree 2 we obtain

$$\begin{aligned}\int_a^b f(x)\,dx &= \int_{x_0}^{x_2} p_2(x)dx + E_2\\ &= \tfrac{1}{6}(b-a)\,[f(x_0)+4f(x_1)+f(x_2)] + E_2,\end{aligned}$$

which is Simpson's rule. Show that

$$E_2 = -\frac{h^5}{90} f^{iv}(\theta), \quad a \le \theta \le b.$$

where $h = \frac{1}{2}(b-a)$.

The trapezium rule will integrate any polynomial of degree ≤ 1 exactly since the error is proportional to f'', whereas Simpson's rule will integrate any cubic without error. In general a quadrature formula is said to have *degree of precision p* if it will integrate all polynomials of degree $\leq p$ without error. The trapezium rule has degree of precision 1, Simpson's rule has degree of precision 3. By introducing further points we can increase the degree of precision further.

Example 5.2.1
The Newton–Cotes *closed* formulae for the integral $\int_a^b f(x)\,dx$ are given by the following. Divide the interval (a, b) into n sub-intervals such that $a = x_0$, $b = x_n$, $x_j = a + jh$, $j = 0, \ldots, n$, $h = (b-a)/n$ then $\int_{x_0}^{x_n} f(x)\,dx$ is given in table 5.2.1. Notice that for n even the degree of precision is $n+1$, but for n odd it is n and this is generally the case (see Isaacson and Keller (1966)). Accordingly, even formulae are usually preferred.

Table 5.2.1 Closed Newton–Cotes formulae.

$n = 1$	$\frac{1}{2}h[f(x_0) + f(x_1)]$
2 point	The trapezium rule: $p = 1 : E_1 = -\frac{1}{12}h^3 f^{ii}(\theta)$
$n = 2$	$\frac{1}{3}h[f(x_0) + 4f(x_1) + f(x_2)]$
3 point	Simpson's $\frac{1}{3}$rd rule: $p = 3$: $E_2 = -\frac{1}{90}h^5 f^{iv}(\theta)$
$n = 3$	$\frac{3}{8}h[f(x_0) + 3f(x_1) + 3f(x_2) + f(x_3)]$
4 point	Simpson's $\frac{3}{8}$th rule: $p = 3 : E_3 = -\frac{3}{80}h^5 f^{iv}(\theta)$
$n = 4$	$\frac{2}{45}h[7f(x_0) + 32f(x_1) + 12f(x_2) + 32f(x_3) + 7f(x_4)]$
5 point	Milne's rule: $p = 5 : E_4 = -\frac{8}{945}h^7 f^{vi}(\theta)$

The formulae given in example 5.2.1 are termed closed since they use the end points of the interval of integration; however, it is useful to define formulae which do not have this property.

Example 5.2.2
The Newton–Cotes *open* formulae are given as follows. We divide the interval (a, b) into $n + 2$ intervals so that $a = x_{-1}$ and $b = x_{n+1}$, i.e. $x_j = a + (j + 1)h$, $j = 0, \ldots, n$, with $h = (b - a)/(n + 2)$; then open quadrature methods for $\int_a^b f(x)\,dx$ are given in table 5.2.2.

Table 5.2.2 Open Newton–Cotes formulae.

$n = 0$	$2hf(x_0)$
1 point	The midpoint rule: $p = 1 : E_0 = \frac{1}{3}h^3 f''(\theta)$
$n = 1$	$\frac{3}{2}h[f(x_0) + f(x_1)]$
2 point	$p = 1 : E_1 = \frac{3}{4}h^3 f''(\theta)$
$n = 2$	$\frac{4}{3}h[2f(x_0) - f(x_1) + 2f(x_2)]$
3 point	$p = 3 : E_2 = \frac{14}{45}h^5 f^{iv}(\theta)$

Example 5.2.3
Consider the integral $\int_0^1 x^{-1/2}\,dx$. It is not possible to apply the closed Newton–Cotes formulae to this integral even though the integral is well defined because the integrand is undefined at $x = 0$. However, there is no such difficulty in applying open formulae which do not require the integrand at $x = 0$.

As the number of points increases we obtain formulae with higher precision but unfortunately the resulting methods become unsuitable. For example, the nine-point closed Newton–Cotes formula is given by

$$\int_{x_0}^{x_9} f(x)\,dx \approx \frac{4h}{14175}[989f(x_0) + 5888f(x_1) - 928f(x_2) + 10496f(x_3)$$
$$-4540f(x_4) + 10496f(x_5) - 928f(x_6) + 5888f(x_7) + 989f(x_8)].$$

This formula has large coefficients which alternate in sign and this can lead to large cancellation errors unless special care is taken.

> 5.3 Gauss quadrature formulae

We shall now attempt to find quadrature formulae of the form

$$\int_{-1}^{1} f(x)\,dx = \sum_{j=0}^{j=n} \alpha_j f(x_j) + E_n, \qquad (5.3.1)$$

which have degree of precision $> n+1$. Such formulae are not restricted to the interval $(-1, 1)$, since given any other interval we can simply map it to this range. (The reason for using this particular range will become clear later.) We will now select the $(2n+2)$ parameters, i.e. the weights, α_j, and the abscissae x_j, $j = 0, \ldots, n$, to make the formula as accurate as possible.

Example 5.3.1 (Analytical)
For $n = 1$ equation (5.3.1) becomes

$$\int_{-1}^{1} f(x)\,dx = \alpha_0 f(x_0) + \alpha_1 f(x_1) + E_1. \tag{5.3.2}$$

As we are looking for a method which has degree of precision > 2 the error E_1 must vanish if $f(x) = 1$. Therefore,

$$\int_{-1}^{1} 1\,dx = 2 = \alpha_0 + \alpha_1. \tag{5.3.3}$$

Similarly, if $f(x) = x$ then $E_1 = 0$ provided

$$\int_{-1}^{1} x\,dx = 0 = \alpha_0 x_0 + \alpha_1 x_1. \tag{5.3.4}$$

We can repeat this for $f(x) = x^2$ to obtain

$$\tfrac{2}{3} = \alpha_0 x_0^2 + \alpha_1 x_1^2 \tag{5.3.5}$$

and for $f(x) = x^3$ to obtain

$$0 = \alpha_0 x_0^3 + \alpha_1 x_1^3. \tag{5.3.6}$$

If all the equations (5.3.3)–(5.3.6) are satisfied then the resulting quadrature formula will be exact for all possible cubics, i.e. it will have degree of precision 3. Notice that the solution of these equations would be identical if we were to integrate from 1 to -1 and so $\alpha_0 = \alpha_1$ and $x_0 = -x_1$ and these conditions enable us to determine

$$\alpha_0 = \alpha_1 = 1, \qquad x_0 = -x_1 = \frac{1}{\sqrt{3}}.$$

This gives the Gauss quadrature formula

$$\int_{-1}^{1} f(x)\,dx = f\left(-\frac{1}{\sqrt{3}}\right) + f\left(\frac{1}{\sqrt{3}}\right) + E_1,$$

where E_1 is proportional to $f^{iv}(\theta)$, $-1 \le \theta \le 1$, since this method has degree of precision 3.

Example 5.3.1 demonstrates the *method of undetermined coefficients* and clearly this approach can be generalized to the Gaussian quadrature formula given by equation (5.3.1). In equation (5.3.1) there are $(2n+2)$ parameters, $(n+1)$ abscissae x_j, $j = 0, \ldots, n$, and $(n+1)$ weights α_j, $j = 0, \ldots, n$. Therefore, we should be able to select the abscissae and weights to give a degree of precision of at least $2n+1$. It is also clear that to use this method for the derivation of Gaussian quadrature formulae will lead to lengthy algebraic manipulation if $n > 1$. Fortunately, there is an alternative approach.

Recall that a set of polynomials, $\phi_j(x)$, $j = 0, \ldots,$ $-1 \leq x \leq 1$, are said to be orthogonal with respect to a weight function $w(x)$ if

$$\int_{-1}^{1} w(x)\phi_j(x)\phi_k(x)\,dx = 0, \qquad j \neq k.$$

We will now show that if we use the $(n+1)$ zeros of $\phi_{n+1}(x)$ as the abscissa in equation (5.3.1) then we can obtain degree of precision $2n+1$. To do this it is necessary to show that $\phi_n(x)$ has exactly n real zeros in the interval $(-1, 1)$. Firstly, ϕ_0 is a constant $C \neq 0$, therefore,

$$0 = \int_{-1}^{1} w(x)\phi_n(x)\phi_0(x)\,dx = C\int_{-1}^{1} w(x)\phi_n(x)\,dx$$

and since $w(x) \geq 0$ this implies that $\phi_n(x)$ has at least one change of sign in $(-1, 1)$ and hence has at least one zero. Recall that ϕ_n is a polynomial of degree n and by the Fundamental Theorem of Algebra has at most n real zeros. Let us assume that the number of real zeros of ϕ_n in $(-1, 1)$ is $j < n$, and that these zeros are at z_i, $i = 1, \ldots, j$, where $-1 < z_1 < z_2 < \ldots < z_j < 1$. With no loss in generality we may assume that $\phi_n > 0$ in $(-1, z_1)$, $\phi_n < 0$ in (z_1, z_2), and so on. Now define $P(x) = (-1)^j \prod_{i=1}^{i=j}(x - z_i)$; then $P(x)\phi_n(x) > 0$, and so

$$\int_{-1}^{1} w(x)P(x)\phi_n\,dx > 0. \qquad (5.3.7)$$

However, P is a polynomial of degree $j < n$ and so it can be expressed as a linear combination

$$P(x) = \sum_{i=0}^{i=j} c_i\phi_i,$$

therefore,

$$\int_{-1}^{1} w(x)P(x)\phi_n \, dx = \sum_{i=0}^{i=j} c_i \int_{-1}^{1} w(x)\phi_i\phi_n \, dx = 0$$

which contradicts (5.3.7). Hence all n zeros of $\phi_n(x)$ lie within $(-1, 1)$.

We have already seen examples of orthogonal polynomials in Chapter 3. In particular the Legendre polynomials are given by $P_0(x) = 1$, $P_1(x) = x$, and

$$(n+1)P_{n+1}(x) = (2n+1)xP_n(x) - nP_{n-1}(x), \quad -1 \leq x \leq 1 \quad (5.3.8)$$

which have weight $w(x) = 1$. From above we see that $P_{n+1}(x)$ has $n+1$ real zeros in the interval $(-1, 1)$, and we will now show that if we select these zeros as the abscissa we obtain a quadrature formula of the form (5.3.1) which has degree of precision $2n+1$. To show this let $P(x)$ be any polynomial of degree $\leq 2n+1$, then there are polynomials $Q(x)$ and $R(x)$ of degree $\leq n$ such that

$$P(x) = Q(x)P_{n+1}(x) + R(x).$$

Furthermore, we can express $Q(x)$ as a linear combination of Legendre polynomials of degree $\leq n$, i.e.

$$Q(x) = \sum_{i=0}^{i=n} d_i P_i(x)$$

hence

$$\int_{-1}^{1} Q(x)P_{n+1}(x) \, dx = \sum_{i=0}^{i=n} d_i \int_{-1}^{1} P_i(x)P_{n+1}(x) \, dx = 0$$

and so

$$\int_{-1}^{1} P(x) \, dx = \int_{-1}^{1} Q(x)P_n(x) \, dx + \int_{-1}^{1} R(x) \, dx = \int_{-1}^{1} R(x) \, dx. \quad (5.3.9)$$

Given any set of abscissae we can select weights α_j such that

$$\int_{-1}^{1} R(x) \, dx = \sum_{j=0}^{j=n} \alpha_j R(x_j) \quad (5.3.10)$$

where the integration is exact since $R(x)$ is a polynomial of degree less than n. Finally, notice that if the interpolation points are selected to be the zeros of $P_{n+1}(x)$ then

$$P(x_j) = Q(x_j)P_{n+1}(x_j) + R(x_j) = 0 + R(x_j) = R(x_j)$$

and so

$$\int_{-1}^{1} P(x)\,dx = \sum_{j=0}^{j=n} \alpha_j P(x_j)$$

with no error, i.e. the quadrature method has degree of precision $2n+1$. It is possible to show[2] that the error of such a formula is given by

$$E_n = \frac{h^{2n+3}((n+1)!)^2}{(2n+3)(2(n+1)!)^3} f^{(2n+2)}(\theta) \tag{5.3.11}$$

where h is the interval width. For the interval $(-1,1)$ we have $h=2$.

Example 5.3.2 (Computational)
In order to illustrate the effectiveness of using the zeros of the Legendre polynomials load the program **QUAD** and then estimate the integral

$$\int_0^1 x^5\,dx$$

using three uniformly spaced points. (This is the default problem for the program **QUAD**.) Press ESCAPE and change the prompt to **Select data** and then press return. Now use a standard grid of three points which are the zeros of the required Legendre polynomial. This produces an estimate for the integral of 0.1666669 compared with the exact value of $\frac{1}{6}$. Recall that in example 5.1.1 it takes six equally spaced points to evaluate this integral exactly; here we have used only three. Notice also that the function which is used to approximate the integrand differs from x^5 by as much as 0.207 and yet the estimate for the integral is very good. The reason for this is given by equation (5.3.11). Using three points we have $n=2$ and so the error term will involve $f^{vi}(\theta)$ which is identically zero for this integrand. (Notice that the program automatically deals with the problem of mapping the zeros of Legendre polynomials, usually defined on the interval ± 1, to a suitable range.)

[2]See Froberg (1987).

Example 5.3.2 (Analytical)
Let us find the Gauss–Legendre quadrature method based on two points, i.e. $n = 1$. The second Legendre polynomial is

$$P_2(x) = \tfrac{1}{2}(3x^2 - 1)$$

which has zeros at $x_0 = -\sqrt{\frac{1}{3}}$ and $x_1 = \sqrt{\frac{1}{3}}$; these points are used as the abscissa. The weight α_0 is given by

$$\begin{aligned}
\alpha_0 &= \int_{-1}^{1} \frac{x - x_1}{x_0 - x_1}\, dx \\
&= \int_{-1}^{1} \frac{(x - \sqrt{\frac{1}{3}})}{-2\sqrt{\frac{1}{3}}}\, dx \\
&= -\frac{1}{2\sqrt{\frac{1}{3}}}\left[\frac{x^2}{2} - x\sqrt{\frac{1}{3}}\right]_{-1}^{1} \\
&= 1
\end{aligned}$$

Similarly $\alpha_1 = 1$. This gives the quadrature formula

$$\int_{-1}^{1} f(x)\, dx \approx f\left(-\frac{1}{\sqrt{3}}\right) + f\left(\frac{1}{\sqrt{3}}\right)$$

which was obtained in example 5.3.1. Finally, the error is given from equation (5.3.11) as $\frac{1}{135}f^{iv}(\theta)$, $|\theta| \leq 1$. We see that this method which is obtained by approximating the integrand at two points has degree of precision 3. Recall that using two equally spaced points the maximum degree of precision is 1.

Example 5.3.3
Values for the weight and absicca of Gauss–Legendre quadrature formulae are given in table 5.3.1.

Example 5.3.4 (Calculator/Computational)
Let us estimate the integral

$$\int_{0}^{1} \exp(-x^2)\, dx$$

Table 5.3.1 Gauss–Legendre quadrature weights and abscissa.

n	x_j	α_j
1	±0.5773502692	1.0000000000
2	0	0.8888888889
	±0.7745966692	0.5555555556
3	±0.3399810436	0.6521451549
	±0.8611363116	0.3478548451
4	0	0.5688888889
	±0.5384693101	0.4786286705
	±0.9061798459	0.2369268851

using Gauss–Legendre quadrature. The range of integration is $(0,1)$ which can be transformed to $(-1,1)$ using the substitution $t = 2x - 1$ but notice that

$$\int_0^1 \exp(-x^2)\,dx = \tfrac{1}{2}\int_{-1}^1 \exp(-x^2)\,dx.$$

Applying the Gaussian quadrature formula with $n = 1$, i.e. two points, gives the result 0.7165315 whilst the exact value is 0.74682413.... This is rather disappointing but the reason for the low accuracy is clear if we sketch the integrand. From figure 5.3.1 the integrand has a relatively large peak at $x = 0$ which is ignored by Gaussian formulae which use an even number of points, i.e. n is odd. However, if we use the Gaussian formula with $n = 4$, i.e. five points we obtain the estimate 0.746832, which is correct to four decimal places. Using the program **QUAD** applied to this integral directly, i.e. without changing the range of integration, and using the zeros of the second Legendre polynomial as abscissa produces an estimate for the integral of 0.7465947. Using the zeros of P_4 produces an estimate of 0.7468245 whereas using four equally spaced points gives 0.7469923.

Exercise 5.3.1 (Computational) Load the program **QUAD** and estimate the integral

$$\int_0^1 \exp(-x^2)\,dx$$

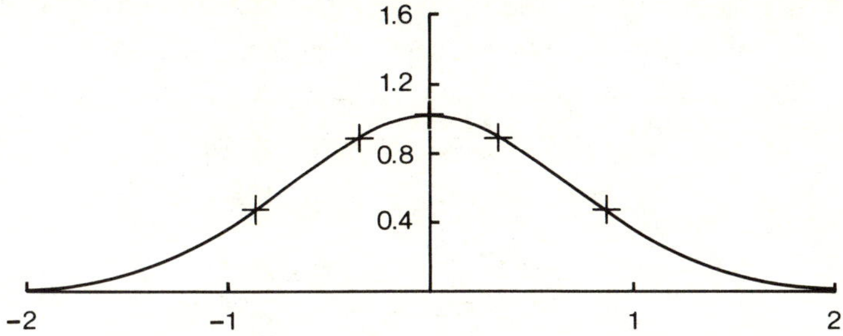

Figure 5.3.1 The integrand $\exp(-x^2)$, $-1 \le x \le 1$.

using two, three, and four equally spaced points. In each case examine the interpolating polynomial which is constructed and compare it with the integrand. Notice that the program computes an error bound between the integrand and the interpolating polynomial and also an estimate of the quadrature formula. Repeat this exercise using the zeros of the corresponding Legendre polynomials.

The Gauss–Legendre formulae were deduced from orthogonal polynomials which are defined only on the interval $(-1, 1)$ and integrals over other ranges must first be reduced to this interval. However, it is frequently the case that other sets of orthogonal polynomials can be used to evaluate integrals defined on other intervals.

Example 5.3.5
Alternative weight functions.

(i) Laguerre polynomials. (See table 5.3.2.) $\int_0^\infty w(x)f(x)\,dx$

$$w(x) = \mathrm{e}^{-x}\colon\ 0 \le x \le \infty,\ L_n(x) = \mathrm{e}^{x}\,(d^n/dx^n)(\mathrm{e}^{-x}x^n).$$

(ii) Hermite polynomials. (See table 5.3.3.) $\int_{-\infty}^{\infty} w(x)f(x)\,dx$

$$w(x) = \exp(-x^2)\colon\ -\infty \le x \le \infty.$$
$$H_n(x) = (-1)^n \exp(x^2)(d^n/dx^n)(\exp(-x^2)).$$

Table 5.3.2 Gauss–Laguerre weights and abscissa.

n	x_j	α_j
1	0.585786	0.853553
	3.414213	0.146447
2	0.415775	0.711093
	2.294280	0.278518
	6.289945	0.010389
3	0.322547	0.603154
	1.745746	0.357419
	4.536620	0.038888
	9.395071	0.000539
4	0.263560	0.521756
	1.413403	0.398667
	3.596426	0.075942
	7.085810	0.003612
	12.640801	0.000023

Table 5.3.3 Gauss–Hermite weights and abscissa.

n	x_j	α_j
1	±0.707107	0.886227
2	0	1.181636
	±1.224745	0.295409
3	±0.524648	0.804914
	±1.650680	0.081313
4	0	0.945309
	±0.958572	0.393619
	±2.0201829	0.019953

> 5.4 Composite quadrature

We have seen that the error associated with simple quadrature is proportional to a power of the interval width. It would seem sensible, therefore, to make the range of integration as small as possible. To do this we write

$$\int_a^b f(x)\,dx = \int_a^{a+h} f(x)\,dx + \int_{a+h}^{a+2h} f(x)\,dx + \ldots + \int_{b-h}^b f(x)\,dx \tag{5.4.1}$$

where $h = (b-a)/N$. We then replace the contribution made in each sub-interval by the corresponding quadrature formula. The result is called a *composite quadrature formula.*

Example 5.4.1 (Composite trapezium rule)
The simple trapezium rule applied to an interval (x_i, x_{i+1}) is given by

$$\int_{x_i}^{x_{i+1}} f(x)\,dx = \tfrac{1}{2}h[f(x_i) + f(x_{i+1})] + E_1^i,$$

where $h = (x_i - x_{i+1})$ and $E_1^i = -\frac{1}{12}h^3 f''(\theta_i)$, $x_i \le \theta_i \le x_{i+1}$. If we set $x_i = a + ih$, with $h = (b-a)/N$, then equation (5.4.1) becomes

$$\begin{aligned}
\int_a^b f(x)\,dx &= \frac{1}{2}h[f(a) + f(a+h)] + E_1^0 \\
&\quad +\frac{1}{2}h[f(a+h) + f(a+2h)] + E_1^1 + \\
&\quad \vdots \\
&\quad +\frac{1}{2}h[f(b-h) + f(b)] + E_1^{N-1} \\
&= \frac{1}{2}h[f(a)+2f(a+h)+2f(a+2h)+\ldots+2f(b-h)+f(b)] \\
&\qquad +E_{\mathrm{CT}},
\end{aligned}$$

where the composite trapezium error, E_{CT}, is given by

$$E_{\mathrm{CT}} = -\sum_{i=0}^{i=N-1} \frac{h^3}{12} f''(\theta_i) = -\frac{Nh^3}{12} f''(\theta).$$

(The mean value theorem for integrals has been used to replace θ_i, $i = 0,\ldots,N-1$, by θ with $a \le \theta \le b$. See Appendix A.) However, $Nh = (b-a)$ and so

$$E_{\mathrm{CT}} = -\tfrac{1}{12}h^2(b-a)f''(\theta). \tag{5.4.2}$$

Notice that the error of the composite trapezium rule is $O(h^2)$ compared with $O(h^3)$ of the simple trapezium rule, but that the h in the composite rule can be made small by introducing more sub-intervals whereas with the latter it is the full interval of integration. By systematically increasing N we should be able to reduce the quadrature error.

Example 5.4.2 (Computational)
Consider the simple trapezium rule for the following integral:

$$\begin{aligned}\int_0^1 \exp(x^2)\,dx &= \tfrac{1}{2}[f(0)+f(1)] + E_1 \\ &= \tfrac{1}{2}[1+2.7182818] + E_1 \\ &= 1.8591409 + E_1.\end{aligned}$$

We can bound the error, E_1, involved in this procedure by using equation (5.4.2), i.e.

$$|E_1| = \left|-\tfrac{1}{12}f''(\theta)\right| \leq \tfrac{1}{12}\,6\exp(1) = \tfrac{1}{2}e^1 = 1.35914$$

since $|f''(x)| = |(4x^2+2)\exp(x^2)| \leq 6$. Therefore,

$$\int_0^1 \exp(x^2)\,dx = 1.8591409 \pm 1.35914.$$

We can improve upon this estimate for $\int_0^1 \exp(x^2)\,dx$ by using the composite trapezium rule. For example, loading the program **INTEGR**, for which the default problem is this inegral, we obtain the results given in table 5.4.1. Notice that the program gives an estimated error bound which is determined by estimating an upper bound for $|f''|$ using finite differences[3]. In all cases this appears to be an overestimate of the quadrature error which is also given in table 5.4.1. The actual error in determining this integral is given in the final column of this table.

Notice that the errors in table 5.4.1 decrease by a factor of four whenever the step size is halved, thus confirming that the error in the composite trapezium rule is proportional to h^2. Clearly, as h tends to zero the value produced by the quadrature formula appears to converge and by systematically reducing h we should obtain an accurate result. Alternatively, we can estimate the number of intervals required to achieve

[3] See Appendix B.

Table 5.4.1 Composite trapezium rule for $\int_0^1 \exp(x^2)\,dx$.

N	Estimate	Estimated error bound	Error bound	Error
2	1.5715832	0.0958526	0.3397852	0.1089310
4	1.4906789	0.0410165	0.0894463	0.0280267
8	1.4697123	0.0143884	0.0212366	0.0070601
16	1.4644203	0.0043409	0.0053091	0.0017681

a given accuracy. For example, let us suppose that we require a final estimate which is correct to six decimal places, i.e. $|E_{\mathrm{CT}}| \leq 0.5 \times 10^{-6}$. To do this we select h such that

$$|E_{\mathrm{CT}}| \leq \tfrac{1}{12}h^2 \max_{0\leq x\leq 1} |f''| \leq 0.5 \times 10^{-6}$$

or

$$\begin{aligned} h^2 &\leq \frac{1}{\mathrm{e}} \times 10^{-6} \\ h &\leq 0.6065 \times 10^{-3}. \end{aligned}$$

From this we conclude that $N = 1/h \geq 1649$ intervals. Using this value for N gives

$$\int_0^1 \exp(x^2)\,dx = 1.462651912 \pm 0.00000017.$$

By increasing N further we might expect to be able to increase the accuracy further but we are limited by the accuracy of the machine being used. To see this recall that any estimate for an integral which is computed by quadrature formulae of this form is determined by the addition of N terms and if each term is evaluated to a precision ϵ then the total has a maximum possible error $N\epsilon$; as N becomes larger the accumulated error will dominate the quadrature error.

Exercise 5.4.1 (Composite Simpson's rule). Show that by replacing an integral by

$$\int_a^b f(x)\,dx = \int_a^{a+2h} f(x)\,dx + \int_{a+2h}^{a+4h} f(x)\,dx + \ldots + \int_{b-2h}^{b} f(x)\,dx$$

where $h = (b-a)/2N$, and then applying Simpson's rule we obtain

$$\int_a^b f(x)\,dx = \tfrac{1}{3}h[f(a) + 4f(a+h) + 2f(a+2h) + 4f(a+3h) + \ldots + 2f(b-2h) + 4f(b-h) + f(b)] + E_{\mathrm{CS}}. \quad (5.4.3)$$

The error of this composite Simpson's rule is $E_{\mathrm{CS}} = -\frac{1}{180}h^4(b-a)f^{iv}(\theta)$, $a \le \theta \le b$.

Exercise 5.4.2 (Computational) Load the program **INTEGR** and estimate the integral $\int_0^1 \exp(-x^2)\,dx$ using the option **Trapezium**. Use the error term E_{CT} to estimate the number of intervals required to determine this integral correct to five decimal places. Repeat this exercise with the composite Simpson rule using the prompt **Simpson's 1/3rd**.

It is clear that we can develop composite integration formulae for any of the higher order Newton–Cotes or Gauss methods. Notice that for hand calculation when using the Newton–Cotes formulae and doubling the number of points from N to $2N$ we can re-use previous calculations. For example with the composite trapezium rule with N intervals we have

$$\int_a^b f(x)\,dx \approx T(h) = \tfrac{1}{2}h[f(a) + 2f(a+h) + \ldots + 2f(b-h) + f(b)]$$

and for $2N$ intervals

$$\begin{aligned} T\left(\tfrac{1}{2}h\right) &= \tfrac{1}{4}h[f(a) + 2f\left(a+\tfrac{1}{2}h\right) + 2f(a+h) \\ &\quad +2f\left(a+\tfrac{3}{2}h\right) + \ldots \\ &\quad +2f(b-h) + 2f\left(b-\tfrac{1}{2}h\right) + f(b)] \\ &= \tfrac{1}{2}T(h) + \\ &\quad \tfrac{1}{2}h\left[f\left(a+\tfrac{1}{2}h\right) + f\left(a+\tfrac{3}{2}h\right) + \ldots + f\left(b-\tfrac{1}{2}h\right)\right]. \end{aligned}$$

Therefore, to find $T(\frac{1}{2}h)$ we need only N more function evaluations than when finding $T(h)$. This is generally the case with Newton–Cotes formulae. The Gaussian quadrature formula with $2N$ points has no abscissa in common with the N-point formula. The composite trapezium method has another advantage; the weights at each abscissa remain the same as the number of intervals is doubled but the same is not true with higher order methods.

Example 5.4.3 (Analytical)
Consider Simpson's rule applied with two intervals to produce

$$\int_a^b f(x)\,dx \approx \tfrac{1}{3}h[f(a)+4f(a+h)+f(b)],$$

where $h=\frac{1}{2}(b-a)$. If we double the number of sub-intervals then

$$\int_a^b f(x)\,dx \approx \frac{1}{6}h\left[f(a)+4f\left(a+\frac{1}{2}h\right)+2f(a+h)+4f\left(a+\frac{3}{2}h\right)+f(b)\right]$$

and we see that the coefficient of $f(a+h)$ has changed from 4 to 2.

Exercise 5.4.3 (Computational) Load the program **INTEGR** and evaluate $\int_0^1 \exp(-x^2)\,dx$ using the options **Gaussian 2 point** and **Gaussian 4 point**. Notice that at each stage you will be asked to specify the number of panels, where the total number of points used will be equal to the number of panels times the number of points in the formula. Compare the results obtained with those in exercise 5.4.2.

Exercise 5.4.4 (Computational) Use the program **INTEGR** to estimate the integral

$$\int_0^1 \frac{1}{\sqrt{x}}\,dx.$$

Why do the trapezium rule and Simpson's rule fail? Compare the results produced by the open Newton–Cotes three-point rule and the Gauss–Legendre methods. Which is the most efficient?

> 5.5 Romberg integration

Whenever we approximate an integral by a quadrature formula it is essential that some estimate of the quadrature error is made. We have seen that using the composite trapezium rule it is possible to show that

$$I=\int_a^b f(x)\,dx = T(h)-\tfrac{1}{12}h^2(b-a)f''(\theta), \qquad a\le x\le b. \qquad (5.5.1)$$

where

$$T(h)=\tfrac{1}{2}h[f(a)+2f(a+h)+\ldots+f(b)].$$

If we halve h then we have

$$I=T\left(\tfrac{1}{2}h\right)-\tfrac{1}{12}\left(\tfrac{1}{2}h\right)^2(b-a)f''(\theta'). \qquad (5.5.2)$$

Now multiply equation (5.5.2) by four and subtract (5.5.1) from the result. We obtain

$$3I = 4T(\tfrac{1}{2}h) - T(h) - \tfrac{1}{12}h^2(b-a)[f''(\theta') - f''(\theta)]. \qquad (5.5.3)$$

If $\theta' \approx \theta$ then

$$\tfrac{1}{3}\left(4T\left(\tfrac{1}{2}h\right) - T(h)\right) = T\left(\tfrac{1}{2}h\right) + \tfrac{1}{3}\left(T\left(\tfrac{1}{2}h\right) - T(h)\right)$$

should be a better approximation for I than either $T(h)$ or $T(\frac{h}{2})$. This process is called *Richardson's method.*

Example 5.5.1 (Computational)
Consider the integral $\int_0^1 \exp(x^2)\,dx$. Using the program **INTEGR** we obtain values for the trapezium rule in table 5.5.1. Let us now apply Richardson's method to the results in the second column. For example, applying equation (5.5.3) with $h = 0.25$ we have

$$\begin{aligned} T(\tfrac{1}{2}h) &+ \tfrac{1}{3}\left[T(\tfrac{1}{2}h) - T(h)\right] \\ &= 1.46971276 + \frac{1.469712276 - 1.490678861}{3} = 1.462723415. \end{aligned}$$

Likewise, we can apply this technique to the remaining entries in the second column of table 5.5.1. Notice that we can achieve 6 correct decimal places using only 32 points instead of the 1649 predicted in example 5.3.1.

Table 5.5.1 Richardson's method applied to $\int_0^1 \exp(x^2)\,dx$.

N	h	Trapezium	Error	Richardson	Error
4	0.25	1.490678861	2.8×10^{-2}		
8	0.125	1.469712276	7.1×10^{-3}	1.462723415	7.2×10^{-5}
16	0.0625	1.464420310	1.8×10^{-3}	1.462656321	4.6×10^{-6}
32	0.03125	1.463094102	4.4×10^{-4}	1.462652033	2.8×10^{-7}
64	0.015625	1.462762347	1.1×10^{-4}	1.462651762	1.8×10^{-8}

Exercise 5.5.1 (Analytical) If $S(h)$ denotes the result of applying Simpson's rule with step h to estimate an integral $I = \int_a^b f(x)\,dx$ then

$$I - S(h) = \frac{h^4}{180}(b-a)\,f^{iv}(\theta), \quad a \le \theta \le b. \qquad (5.5.4)$$

Show that

$$\frac{16S(\frac{1}{2}h) - S(h)}{15} = S(\tfrac{1}{2}h) + \frac{[S(\frac{1}{2}h) - S(h)]}{15} \qquad (5.5.5)$$

gives a better approximation for I.

Richardson's method can be applied in a similar way to any other quadrature rule but it can be developed to produce a systematic extrapolation procedure. Let us assume that $I(h)$ is a result produced by some quadrature rule when applied to an integral I and that the error associated with this approximation is given by

$$I = I(h) + Ah^p + \mathrm{O}(h^{p+1}) \qquad (5.5.6)$$

where A is some constant independent of h. Similarly,

$$I = I(\tfrac{1}{2}h) + A(\tfrac{1}{2}h)^p + \mathrm{O}((\tfrac{1}{2}h)^{p+1}). \qquad (5.5.7)$$

Now we multiply (5.5.7) by 2^p and then subtract (5.5.6) from the result. This produces

$$(2^p - 1)I = 2^p I(\tfrac{1}{2}h) - I(h) + \mathrm{O}(h^{p+1}) \qquad (5.5.8)$$

$$I = \frac{2^p I(\frac{1}{2}h) - I(h)}{2^p - 1} + \mathrm{O}(h^{p+1}). \qquad (5.5.9)$$

Therefore, if $I(h)$ and $I(\frac{1}{2}h)$ are of precision $p-1$ then

$$\frac{2^p I(\frac{1}{2}h) - I(h)}{2^p - 1} = I(\tfrac{1}{2}h) + \frac{I(\frac{1}{2}h) - I(h)}{2^p - 1} \qquad (5.5.10)$$

has degree of precision $\geq p$. This process can be repeated and is called *Romberg extrapolation.*

Example 5.5.2 (Analytical)
It can be shown that if $T(h)$ is the result of applying the trapezium rule to estimate an integral $I = \int_a^b f(x)\,dx$ then

$$I = T(h) + Ah^2 + Bh^4 + Ch^6 + \ldots \qquad (5.5.11)$$

where A, B, C, ..., are *constants independent of* h. These constants depend only upon f in the sense that A is related to f'', B to f^{iv} and

Table 5.5.2 Romberg extrapolation applied to $\int_0^1 \exp(x^2)\,dx$.

N	h	$T(h)$	$T(h, \frac{1}{2}h)$	$T(h, \frac{1}{2}h, \frac{1}{4}h)$
4	0.25	1.490678861		
8	0.125	1.469712276	1.462723415	
16	0.0625	1.464420310	1.462656321	1.462651849
32	0.03125	1.463094102	1.462652033	1.462651747

so on. Therefore, provided f has a bounded second derivative we can eliminate the term Ah^2 to obtain equation (5.5.10) with $p = 2$, i.e.

$$T(h, \tfrac{1}{2}h) = T(\tfrac{1}{2}h) + \tfrac{1}{3}\left[T(\tfrac{1}{2}h) - T(h)\right],$$

which has error proportional to h^4. From this we have that

$$I = T(h, \tfrac{1}{2}h) + B'h^4 + C'h^6 + \ldots$$

where B' is independent of h but does depend on the fourth derivative of f. If f has a bounded fourth derivative we can apply (5.5.10) to eliminate the term $B'h^4$ and so on. As an example let us use the program **INTEGR** to evaluate $\int_0^1 \exp(x^2)\,dx$. This particular integral is set up as default, therefore, select the option **Romberg** and start with $h = 0.5$ to produce the Romberg tableau shown in table 5.5.2. Notice that we look down the columns for convergence because the terms across the rows must necessarily approach one another by the way in which they are defined.

Although Romberg extrapolation can produce a dramatic increase in the accuracy of a quadrature rule it must be used with care.

Example 5.5.3 (Analytical)
Consider the integral $\int_0^1 \sqrt{x}\log(x)\,dx$. The integrand is shown in figure 5.5.1 and is clearly defined at $x = 0$ but this function does not have a bounded second derivative throughout the interval and so the expression (5.5.11) is not valid. This integral can be carried out by making a change of variable, $x = t^2$, which transforms the integral to $\int_0^1 4t^2\log(t)\,dt$, the integrand of which has bounded derivatives of all orders.

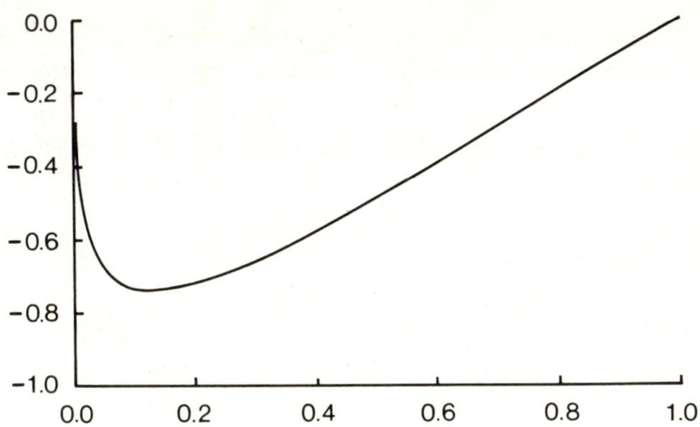

Figure 5.5.1 The integrand $\sqrt{x}\log(x)$.

Romberg extrapolation is based on the existence of the expression (5.5.11), but even if such an expression is not valid it is sometimes possible to find an expression for the error. For example, for integrals of the form

$$I = \int_0^1 x^{1/2}\log(x)g(x)\,dx,$$

where $g(x)$ is sufficiently smooth, the error in the trapezium rule, $T(h)$, is

$$I - T(h) = Ah^{\frac{3}{2}}\log(h) + Bh^{\frac{3}{2}} + Ch^2 + Dh^{\frac{5}{2}}\log(h) + Eh^{\frac{5}{2}} + \ldots.$$

Given a sequence of values $h_1, h_2, h_3, \ldots$ the terms in this series can be eliminated systematically. (See Fox and Mayers (1965).)

Exercise 5.5.2 (Computational) Use the option **Romberg** in the program **INTEGR** to evaluate $\int_0^1 x^7\,dx$. Why do three extrapolations produce an exact answer?

Exercise 5.5.3 (Analytical) If $T(h)$ and $S(h)$ are results produced by the trapezium and Simpson's method show that

$$S(\tfrac{1}{2}h) = T(\tfrac{1}{2}h) + \tfrac{1}{3}\left[T(\tfrac{1}{2}h) - T(h)\right].$$

Furthermore, show that the first extrapolation of Simpson's rule, and therefore, the second extrapolation of the trapezium rule, produces the closed Newton–Cotes rule for four points given in table 5.2.1. (The resulting method is called Simpson's $\frac{3}{8}$th rule.)

Exercise 5.5.4 (Computational) Use the program **INTEGR** with option **Trapezium** to estimate the integral $\int_0^1 \exp(x^2)\,dx$. Repeat this exercise for the prompts **Simpson's 1/3rd** and **Simpson's 3/8th**. Confirm that these values are the first, second and third columns of the corresponding Romberg table.

> 5.6 Adaptive integration

With all quadrature formulae the choice of the step size, h, is critical. Too large a value and the result will be poor, too small a value and the amount of work required will be large and the result may be prone to rounding errors. In this section we will now outline an procedure which can automatically adjust the step size. To see the importance of this approach consider the following example.

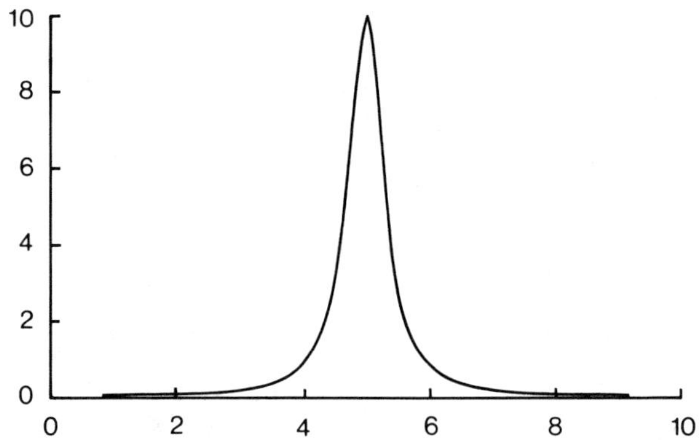

Figure 5.6.1 The integrand $[(x-5)^2+0.1]^{-1}$.

Example 5.6.1 (Analytical)
Let us estimate the integral

$$\int_0^{10} \frac{1}{(x-5)^2+0.1}\,dx.$$

The integrand is shown in figure 5.6.1. If we were to apply Simpson's rule with a uniform step size h to produce an estimate $S(h)$ then the error involved is proportional to $h^4 f^{iv}$. Throughout most of the interval f^{iv} will be very small; only in the neighbourhood of $x = 5$ will it be significant. However, to ensure that we achieve an overall precision ϵ we need to select h such that

$$\left|\frac{h^4 f^{iv}}{180}\right| \leq \epsilon$$

i.e.

$$h \leq \left(\frac{180\epsilon}{|f^{iv}|_{\max}}\right)^{1/4}$$

in which case we will be forced to use a small value for h even though the fourth derivative is very small except near $x = 5$. In order to overcome this difficulty we will now permit the sub-division of the range of integration and use non-uniformly spaced abscissa. Let $I_a^b(h)$ be an approximation for the integral $I = \int_a^b f(x)\,dx$ which has degree of precision p, i.e.

$$I - I_a^b(h) = Ah^{p+1}\, f^{(p+1)}(\theta). \qquad (5.6.1)$$

If we require a final accuracy ϵ then we should select h so that

$$h \leq \left(\frac{\epsilon}{|Af^{(p+1)}|_{\max}}\right)^{1/p}$$

but in most cases $|f^{(p+1)}|_{\max}$ is unknown. In order to overcome this we sub-divide the range of integration (a, b) into (a, c) and (c, b), where $c = \frac{1}{2}(a + b)$, then

$$\begin{aligned} I_a^c - I_a^c(\tfrac{1}{2}h) &= A\left(\tfrac{1}{2}h\right)^{p+1} f^{(p+1)}(\theta_1) \\ I_c^b - I_b^c(\tfrac{1}{2}h) &= A\left(\tfrac{1}{2}h\right)^{p+1} f^{(p+1)}(\theta_2), \end{aligned}$$

where $a \leq \theta_1 \leq c$ and $c \leq \theta_2 \leq b$. Therefore, there will exist a value θ' in (a, b) such that

$$I - I_a^c(\tfrac{1}{2}h) - I_c^b(\tfrac{1}{2}h) = \frac{h^{p+1}}{2^p} A f^{(p+1)}(\theta').$$

If we subtract equation (5.6.1) from this then

$$I_a^b(h) - I_a^c(\tfrac{1}{2}h) - I_c^b(\tfrac{1}{2}h) \approx Ah^{p+1} f^{(p+1)}(\theta)(1 - 2^{-p})$$

provided $\theta \approx \theta'$. This last equation enables us to estimate $f^{(p+1)}(\theta)$, or more precisely

$$Ah^{p+1}f^{(p+1)}(\theta) = \frac{(I_a^b(h) - (I_a^c(\frac{1}{2}h) + I_c^b(\frac{1}{2}h)))}{(1-2^{-p})}.$$

Hence, if

$$|I_a^b(h) - I_a^c(\tfrac{1}{2}h) - I_c^b(\tfrac{1}{2}h)| \leq (2^p - 1)\epsilon$$

then

$$|I - (I_a^c(\tfrac{1}{2}h) + I_c^b(\tfrac{1}{2}h))| \leq \epsilon$$

as required. If not we reconsider each sub-interval independently and use a criterion of $\frac{1}{2}\epsilon$ in each sub-interval.

Example 5.6.2 (Computational)
If we use the program **INTEGR** with the option **Simpson 1/3rd** to evaluate the integral

$$\int_0^{10} \frac{1}{(x-5)^2 + 0.1}\,dx$$

then 512 intervals are required to ensure that two successive estimates agree to eight decimal places. Using the option **Gaussian 4 point Rule** 128 panels of four points are needed to achieve the same accuracy and so there is little to be gained in using this method. However, using the adaptive procedure with a relative error tolerance of 10^{-8} only 161 function evaluations are required.

> 5.7 The Clenshaw–Curtis method

We have seen in the previous sections how it is possible to estimate an integral by approximating the integrand by a suitable polynomial, integrate the approximation and then estimate the error. In order to obtain the best polynomial approximation, of a given order, it would seem obvious to use the minimax approximation. However to do this we need to expend more effort in finding the minimax approximation than we would use with a poorer approximation of higher order. As a compromise we could use an approximation which is almost minimax but which can be found a great deal easily, for example a Chebyshev approximation. This observation is the basis of the Clenshaw–Curtis method.

Let us assume that we have a Chebyshev approximation for a function $f(x)$ which is given by

$$f(x) = \tfrac{1}{2}c_0T_0 + c_1T_1(x) + c_2T_2(x) + \ldots, \qquad (5.7.1)$$

then the indefinite integral of $f(x)$ is given by

$$\int f(x)\,dx = \tfrac{1}{2}c_0 \int T_0\,dx + c_1 \int T_1(x)\,dx + \ldots. \qquad (5.7.2)$$

However,

$$\begin{aligned} \int T_0\,dx &= \int 1\,dx = x + C_0 = T_1(x) + C_0, \\ \int T_1\,dx &= \int x\,dx = \tfrac{1}{2}x^2 + C_1 = \tfrac{1}{4}T_2(x) + C_1', \end{aligned}$$

where C_0, C_1 and C_1' are constants of integration, and

$$\int T_n(x)\,dx = \frac{1}{2}\left(\frac{T_{n+1}}{n+1} - \frac{T_{n-1}}{n-1}\right), \quad n \geq 2. \qquad (5.7.3)$$

This last result can be obtained as follows:

$$\begin{aligned} \int T_n(x)\,dx &= \int \cos(n\cos^{-1}x)\,dx \\ &= \int -\cos(n\theta)\sin\theta\,d\theta, \quad (x = \cos\theta) \\ &= -\tfrac{1}{2}\int [\sin(n+1)\theta - \sin(n-1)\theta]\,d\theta \\ &= \frac{1}{2}\left(\frac{T_{n+1}}{n+1} - \frac{T_{n-1}}{n-1}\right). \end{aligned}$$

We now recall that T_{2n} is an even function and that T_{2n+1} is an odd function and then use these results to show that

$$[T_{2n+1}]_{-1}^{1} = 2, \text{ and } [T_{2n}]_{-1}^{1} = 0.$$

Substituting these expressions into equation (5.7.2) we determine

$$\begin{aligned} \int_{-1}^{1} f(x)\,dx &= 2\left[\frac{c_0}{2} + \frac{1}{2}c_2\left(\frac{1-3}{1\times 3}\right) + \frac{1}{2}c_4\left(\frac{3-5}{3\times 5}\right) + \ldots\right] \\ &= 2\left(\frac{c_0}{2} - \frac{c_2}{1\times 3} - \frac{c_4}{3\times 5} - \ldots\right) \\ &= c_0 - \sum_{r=2}^{\infty} \frac{1+(-1)^r}{r^2-1}c_r \end{aligned}$$

and if we knew the Chebyshev coefficients of the function $f(x)$ we could find the required integral. Notice that the above series will converge much faster than the Chebyshev series for the integrand because of the term $r^2 - 1$ in the denominator. Unfortunately, the determination of the Chebyshev coefficients is time consuming and in general will involve integrals of the form

$$c_n = \frac{2}{\pi} \int_0^{\pi} \cos(n\theta) f(\cos\theta)\, d\theta \tag{5.7.4}$$

which may be as complicated as the original integral. However, for the moment let us assume that we apply the composite trapezium rule with N points to the integral (5.7.4) then we find approximate coefficients c'_n as

$$c'_n = \frac{2}{N}\left(\tfrac{1}{2}g_0 + g_1 + \ldots + g_{N-1} + \tfrac{1}{2}g_N\right), \tag{5.7.5}$$

where $g_k = \cos(kn\pi/N)\, f(\cos(k\pi/N))$. Notice that in this form we can take advantage of the fact that when the number of points is doubled from N to $2N$ we only need to evaluate g at a further N points. It is possible to show that

$$c'_n = c_n + c_{2N-n} + c_{2N+n} + c_{4N-n} + c_{4N+n} \tag{5.7.6}$$

and if the coefficients of the Chebyshev series approach zero rapidly then the approximate coefficients should produce a good approximation.

Example 5.7.1 (Computational)
Consider the integral

$$\int_{-1}^{1} \exp(x^2)\, dx$$

for which the results in table 5.7.1 were obtained. From table 5.7.1 we see that provided $N > n$ we obtain a very good approximation for the real coefficient which is given in the final column. If $N = n$ then the approximation is out by a factor of 2, which can be confirmed by putting $N = n$ in (5.7.6). We can now use these results to estimate the required integral. For example, if we include only the first four terms of the series given by equation (5.7.2) then

$$\int_{-1}^{1} \exp(x^2)\, dx \approx 2\left(\frac{c_0}{2} - \frac{c_2}{3} - \frac{c_4}{15} - \frac{c_6}{35}\right) = 2.92532116,$$

where the first neglected term is $c_8/63 \approx 1.746 \times 10^{-5}$. The exact value of this integral is 2.9253035 and the above estimate has error 1.76×10^{-5}.

Table 5.7.1 The Chebyshev coefficients of $\exp(x^2)$.

N	2	4	8	16	Exact
$n = 0$	3.7182818	3.5078622	3.5067753	3.5067753	3.5067753
$n = 2$	1.7182818	0.8591409	0.8503917	0.8503917	0.8503917
$n = 4$	3.7182818	0.2104196	0.1052098	0.1052087	0.1052087
$n = 6$	1.7182818	0.8591410	0.0087492	0.0087221	0.0087221
$n = 8$	3.7182818	3.5078622	0.0010869	0.0005434	0.0005434

Exercise 5.7.1 (Computational) Use the option **Trapezium** in the program **INTEGR** to determine the first six Chebyshev coefficients for the function $\exp(-x^2)$. How many function evaluations are required to determine the integral

$$\int_{-1}^{1} \exp(-x^2)\, dx$$

correct to four decimal places using the method of Clenshaw and Curtis? If the trapezium rule were applied directly to this integral how many intervals are required to determine a result which is correct to four decimal places? Which method is the most efficient and why?

> Chapter 6

> Differential Equations: Boundary Value Problems

'Pinning it down at both ends?'

> 6.1 Introduction

In this chapter we shall discuss ways in which it is possible to approximate the solution of differential equations of the form

$$y'' = f(x, y, y'). \tag{6.1.1}$$

The general solution of this equation involves exactly two arbitrary constants. If two additional conditions are supplied then we may be able to find a unique solution, however, it is possible that the conditions which are supplied are not satisfied by any member of the class of functions determined by the general solution. If both conditions are specified at the same point then the problem is said to be an *initial value problem* and in this case under relatively simple conditions it is possible to tell whether or not a particular problem has a unique solution. If the two conditions are specified at different points then the problem is called a *boundary value problem*. There are relatively few cases when it is possible to say that a particular boundary value problem has a unique solution. (For a simple introduction see Harding and Quinney (1986).) Because of the difficulties involved with boundary value problems we will concentrate on linear problems which may be written in the form

$$y'' + p(x)y' + q(x)y = r(x), \tag{6.1.2}$$

$$y(a) = \alpha, \quad y(b) = \beta, \qquad a \le x \le b.$$

Without loss of generality we may take $a = 0$ and $b = 1$ for if this is not the case we can always transform any finite interval (a, b) to $(0, 1)$ by means of a suitable mapping. We will also use the interval $(-1, 1)$ when it is more convenient to do so.

In *A Simple Introduction to Numerical Analysis* it was shown how it was possible to determine an approximate solution of the two-point boundary value problem (6.1.2) by either a *shooting method* or by setting up a system of linear equations produced by replacing the derivatives in the differential equation by finite differences. In either case we obtain a representation of the solution only at a finite set of points; if we require the solution at a non-mesh point then we have to use some form of interpolation. In this chapter we will see how it is possible to obtain a solution in the form

$$y_N(x) = \sum_{i=0}^{i=N} c_i \, \phi_i(x), \tag{6.1.3}$$

where each of the functions $\phi_i(x)$ is defined for all points in the interval in which the solution is required.

> 6.2 Solution in series

We begin by considering equations of the form

$$y'' + p(x)y' + q(x)y = r(x), \qquad 0 \le x \le 1, \tag{6.2.1}$$

where p and q are rational functions. If p and q are continuous functions at a point x_0 then x_0 is said to be an *ordinary point*. If p and q are not continuous at x_0, but $(x - x_0)p(x)$ and $(x - x_0)^2 q(x)$ are continuous at x_0 then x_0 is said to be a *regular singularity*. All other points are called irregular singularities. Near an ordinary point the solution of the differential equation can be expressed as a Taylor series expansion which converges as far as the next singularity. Near a regular singularity the solutions y_1 and y_2 are of the form

$$y_1(x) = (x - x_0)^{\rho_1} \sum_{n=0}^{n=\infty} c_n (x - x_0)^n$$

and

$$y_2(x) = C \log(x - x_0) y_1(x) + (x - x_0)^{\rho_2} \sum_{n=0}^{n=\infty} d_n (x - x_0)^n,$$

where ρ_1, ρ_2 and C, together with c_n and d_n, are to be determined. In both cases the series converges as far as the next discontinuity of either p or q. If $C \neq 0$ then these solutions are linearly independent. However, it is possible to show that in all cases the general solution of the differential equation is given by

$$y(x) = c_1 y_1(x) + c_2 y_2(x).$$

Since such solutions are known to exist we can substitute these expressions into the differential equation and try to determine values for ρ_1, ρ_2, C, and the coefficients c_n and d_n, $n = 0, 1, \ldots$. This is usually called *Frobenius's method* or the method of solution in series. (See Bender and Orszag (1978).)

Example 6.2.1 (Analytical)
The differential equation

$$x^2 y'' + 2y = 0$$

has a regular singularity at $x_0 = 0$, therefore, we look for solutions of the form

$$y(x) = x^\rho \sum_{n=0}^{n=\infty} c_n x^n.$$

Substituting this expression into the differential equation we find that

$$x^2 y'' + 2y = c_0(\rho(\rho - 1) + 2)x^\rho + c_1(\rho^2 + \rho + 2)x^{\rho+1} + \ldots.$$

However, the right-hand side must vanish identically, and so we select the coefficients and ρ so that this is the case. The coefficient of x^ρ is

$$c_0(\rho^2 - \rho + 2)$$

which is zero for all values of c_0 provided $\rho = 1$ or $\rho = 2$. The coefficient of $x^{\rho+1}$ is $c_1(\rho^2 + \rho + 1)$ and since $\rho^2 + \rho + 1 \neq 0$ for either ρ_1 or $\rho = 2$ we must have $c_1 = 0$. Furthermore, for these choices of ρ all the other coefficients are zero and so we obtain two solutions $y_1(x) = x$ and $y_2(x) = x^2$. The general solution is then given by

$$y(x) = Ax + Bx^2,$$

where A and B are arbitrary constants. Notice that even though the differential equation has a regular singularity at $x = 0$ the general solution is a continuous function near this point.

Exercise 6.2.1 (Analytical) Use the method of Frobenius to determine solutions of

$$xy'' - xy' + y = 0,$$

which are valid for all values $|x| \leq \infty$.

Example 6.2.1 is a particularly simple example and in general it will be far more difficult to derive a simple expression for the solution. However, it does suggest that whenever we are looking for a continuous solution we should consider possible solutions in terms of a simple Taylor series expansion, i.e. solutions of the form

$$y(x) = \sum_{n=0}^{n=\infty} c_n x^n. \tag{6.2.2}$$

This type of expansion was considered in Chapter 3 and we have seen that in most cases it is more convenient to consider a Chebyshev expansion instead, therefore, we now look for solutions of the form

$$y(x) = \sum_{n=0}^{n=\infty}{}' c_n T_n(x). \tag{6.2.3}$$

If we were to substitute equation (6.2.3) directly into the differential equation

$$y'' + p(x)y' + q(x)y = r(x), \tag{6.2.4}$$

then we would need to differentiate each Chebyshev polynomial which requires considerable effort. Instead we integrate the differential equation (6.2.4) to give

$$y' + py + \int (-p' + q)y \, dx = \int r \, dx + A$$

and then again to produce

$$y + \int py \, dx \int\int (-p' + q)y \, dx dx = \int\int r \, dx dx + Ax + B,$$

where A and B are arbitrary constants. We can now use some more properties of Chebyshev series to simplify this expression. In particular

$$\int T_n(x) \, dx = \frac{T_{n+1}}{2(n+1)} - \frac{T_{n-1}}{2(n-1)}, \quad n \geq 2 \tag{6.2.5}$$

with

$$\int T_0(x)\,dx = \int 1\,dx = x = T_1(x),$$

and

$$\int T_1(x)\,dx = \int x\,dx = \tfrac{1}{2}x^2 = \tfrac{1}{4}(T_2(x) + T_0(x)).$$

(See equation (5.7.3).) Also

$$x^r T_s(x) = \frac{1}{2^r}\sum_{i=0}^{i=r} \binom{r}{i} T_{s-r+2i}(x). \qquad (6.2.6)$$

This last expression appears very complicated but it provides a key to determining the solution of the differential equation (6.2.4).

Exercise 6.2.1 (Analytical) Deduce the expression (6.2.6). (Hint: recall that the Chebyshev polynomials are defined by

$$T_{n+1}(x) = 2xT_n(x) - T_{n-1}(x)$$

which may be rearranged to give

$$xT_n(x) = \tfrac{1}{2}(T_{n+1}(x) + T_{n-1}(x))$$

and this is equation (6.2.6) with $r = 1$. Now write $x^r T_s = x^{r-1}(xT_s)$.)

Example 6.2.2 (Analytical/Computational)
Consider the solution of the differential equation

$$y'' - y = 1 + x, \qquad -1 \le x \le 1,$$

subject to the conditions $y(0) = 1$, $y(1) = 3$. Integrating the differential equation twice gives

$$y + \int\!\!\int y\,dx\,dx = A + Bx + \tfrac{1}{2}x^2 + \tfrac{1}{6}x^3.$$

Substituting equation (6.2.3) the integral contains terms $\int T_n\,dx$ which can be replaced using equation (6.2.5). Next we replace the polynomial on the right-hand side of the differential equation by the equivalent Chebyshev series. For example,

$$\begin{aligned} \tfrac{1}{2}x^2 &= \tfrac{1}{4}(T_2 + T_0), \\ \tfrac{1}{6}x^3 &= \tfrac{1}{24}(T_3 + T_1). \end{aligned}$$

Finally, we collect together all the coefficients of each Chebyshev polynomial, T_n, for $n \geq 2$. The coefficients of T_0 and T_1 involve the arbitrary constants A and B and are therefore omitted. This gives

$$\begin{aligned}
c_2 + \left(\tfrac{c_0-c_2}{2\times 4} - \tfrac{c_2-c_4}{4\times 6}\right) &= \tfrac{1}{4} \\
c_3 + \left(\tfrac{c_1-c_3}{4\times 6} - \tfrac{c_3-c_5}{6\times 8}\right) &= \tfrac{1}{24} \\
c_4 + \left(\tfrac{c_2-c_4}{6\times 8} - \tfrac{c_4-c_6}{8\times 10}\right) &= 0 \\
c_5 + \left(\tfrac{c_3-c_5}{8\times 10} - \tfrac{c_5-c_7}{10\times 12}\right) &= 0 \\
\vdots \quad &= \quad \vdots
\end{aligned}$$

At $x = 0$ we use the boundary condition $y(0) = 1$ with

$$T_n(0) = \begin{cases} 0, & n \text{ odd}, \\ (-1)^{n/2}, & n \text{ even}, \end{cases}$$

and so

$$y(0) = \tfrac{1}{2}c_0 - c_2 + c_4 - c_6 + \ldots = 1.$$

Similarly, at $x = 1$

$$y(1) = \tfrac{1}{2}c_0 + c_1 + c_2 + c_3 + \ldots = 3.$$

Therefore, we have an infinite set of linear equations in the unknowns $c_0, c_1, \ldots$, but as we expect them to tend to zero rapidly let us take the equations which come from the boundary conditions and the first four which are derived from the differential equation. All other coefficients are assumed to vanish. The result is a system of six linear equations in c_0, c_1, c_2, c_3, c_4 and c_5. Solving the resulting equations by Gaussian elimination with partial pivoting gives

$$\begin{aligned}
c_0 &= 2, & c_1 &= 2.045904223, \\
c_2 &= 0, & c_3 &= -0.04649799, \\
c_4 &= 0, & c_5 &= 0.000593589.
\end{aligned}$$

The solution of this problem is $y(x) = 1 + x + \frac{\sin(x)}{\sin(1)}$ which has the Chebyshev expansion

$$y(x) = 1 + 2.0459079T_1 - 0.0464980T_3 + 0.0005936T_5 - 0.00000357T_7.$$

Example 6.2.2 demonstrates how it is possible to use Chebyshev polynomials to produce approximate solutions of a limited class of linear differential equations. This method can be quite advantageous if more general boundary conditions are supplied.

Exercise 6.2.2 (Analytical/Computational) Use Chebyshev polynomials to solve the differential equation

$$y'' + xy' + xy = 1 + x + x^2,$$

subject to the boundary conditions

$$y(0) = 1, \qquad y'(0) + 2y(1) - y(-1) = -1.$$

The boundary conditions here are very complicated but the solution of the problem is very similar to that in example 6.2.1. (Hint: terms such as $x^r T_n(x)$ can be replaced using equation (6.2.6) and integrals such as $\int T_n(x)\,dx$ can be written as $\frac{1}{2}\int (T_{n+1} + T_{n-1})\,dx$.)

> 6.3 Finite element methods

Both Frobenius's method and the Chebyshev expansion assume an expansion of the solution as an infinite series. We will now consider a similar approach based on finite expansions. For the sake of clarity we will consider only linear differential equation which are written in the self-adjoint form

$$(p(x)y')' + q(x)y = f(x), \tag{6.3.1}$$

subject to the homogeneous boundary conditions $y(0) = y(1) = 0$. It is relatively easy to show that any linear two-point boundary value problem subject to known boundary values can be reduced to this form.

Example 6.3.1 (Analytical)
As a simple example let us reduce the equation

$$xz'' + z' + x^3 z = x - 1, \quad z(0) = 0, \quad z(1) = 1,$$

to the self-adjoint form (6.3.1). The first step is to reduce the boundary conditions to a homogeneous form. To do this set $y(x) = z(x) - x$ for then $y(0) = z(0) - 0 = 0$ and $y(1) = z(1) - 1 = 0$ as required. Similarly, the differential equation becomes

$$xy'' + y' + x^3 y = -1 + x - x^4.$$

If we rewrite equation (6.3.1) as $py'' + p'y' + qy = f$ and compare coefficients then we see that $p' = 1$ and so $p = x$. Similarly, $q(x) = x^3$ and $f(x) = -1 + x - x^4$, and so the given equation becomes

$$(xy')' + x^3 y = -1 + x - x^4,$$

subject to $y(0) = y(1) = 0$.

Exercise 6.3.1 (Analytical) Reduce the problem

$$(1+x)^2 z'' = 2z - 4, \qquad z(0) = 0, \quad z(1) = 1$$

to the self-adjoint form (6.3.1).

The idea of the finite element method is to select a set of linearly independent basis functions $\phi_i(x)$, $i = 1, \ldots, N$, each of which satisfies the homogeneous boundary conditions

$$\phi_i(0) = \phi_i(1) = 0$$

and then to look for an approximate solution of the differential equation of the form

$$Y_N(x) = \sum_{i=1}^{i=N} c_i \phi_i(x). \tag{6.3.2}$$

Notice that $Y_N(x)$ is defined at every point of the interval $(0, 1)$ and automatically satisfies the boundary conditions. The problem now is to decide which basis function to use and then how to determine the coefficients c_i, $i = 1, .., N$, in such a way as to make $Y_N(x)$ the best possible approximation for the solution, i.e. we wish to minimize the error between $y(x)$ and $Y_N(x)$.

Example 6.3.2
Two possible sets of basis functions are

1. $\phi_i(x) = \sin(i\pi x)$, $i = 1, \ldots, N$, $0 \le x \le 1$,

2. $\phi_i(x) = x(1-x)x^{i-1}$, $i = 1, \ldots, N$, $0 \le x \le 1$.

The problem of determining suitable values for the coefficients c_i may be tackled in a variety of different ways.

> 6.3.1 The method of collocation

Differentiating equation (6.3.2) and substituting the resulting expression into equation (6.3.1) we obtain

$$p(x)\sum_{i=1}^{i=N} c_i\phi_i''(x) + p'(x)\sum_{i=1}^{i=N} c_i\phi_i'(x) + q(x)\sum_{i=1}^{i=N} c_i\phi_i(x) = f(x)$$

or

$$\sum_{i=1}^{i=N}[p(x)\phi_i''(x) + p'(x)\phi_i'(x) + q(x)\phi_i(x)]c_i = f(x).$$

If $Y_N(x)$ were the exact solution then this expression would be satisfied identically, but, since we have only N parameters at our disposal, i.e. c_i, $i = 1, \ldots, N$, we cannot expect it to be satisfied at more than N points. We can select any N points, x_k, $k = 1, \ldots, N$, but once they are specified the solution can be determined by solving the linear equations

$$\sum_{i=1}^{i=N}[p(x_k)\phi_i''(x_k)+p'(x_k)\phi_i'(x_k)+q(x_k)\phi_i(x_k)]c_i = f(x_k),\ k = 1, \ldots, N, \tag{6.3.3}$$

by Gaussian elimination with partial pivoting, if necessary. This is called a *collocation method* and the points x_k, $k = 1, \ldots, N$, are called the *collocation points.*

Example 6.3.3 (Computational)
In order to demonstrate the collocation method let us take the basis function $\phi_i(x) = \sin(i\pi x)$, $i = 1, 2, 3$, and then use the collocation points $x_k = 0.25, 0.5, 0.75$, to compute an approximate solution for the boundary value problem

$$y'' - \frac{2}{(1+x)^2}y = \frac{(2x-4)}{(1+x)^2}, \qquad y(0) = y(1) = 0.$$

The trial solution is given by

$$Y_3(x) = c_1 \sin \pi x + c_2 \sin 2\pi x + c_3 \sin 3\pi x,$$

where the coefficients c_1, c_2 and c_3 satisfy the linear equations

$$\mathbf{Ac} = \mathbf{b},$$

with the components of **A** given by

$$a_{ki} = \left(-(i\pi)^2 - \frac{2}{(1+x_k)^2}\right) \sin i\pi x_k$$

and those of **b** by

$$b_k = \frac{(2x_k - 4)}{(1+x_k)^2}, \qquad k = 1, 2, 3.$$

Solving these equations gives

$$c_1 = 0.161463, \quad c_2 = 0.016691, \quad c_3 = 0.0045005.$$

This trial solution is compared with the exact solution of this problem, $y(x) = x(1-x)(1+x)^{-1}$, in figure 6.3.1.

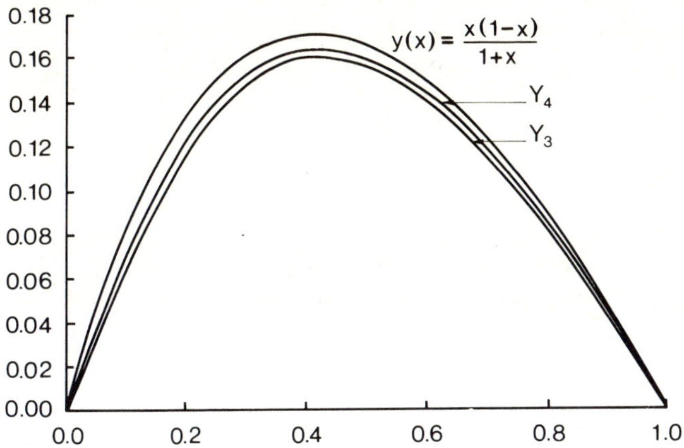

Figure 6.3.1 Trial solutions Y_3 and Y_4 for example 6.3.3.

Exercise 6.3.2 (Computational) Load the program **FEM** for which the default problem is the solution of the two-point boundary value problem

$$y'' - \frac{2}{(1+x)^2}y = \frac{(2x-4)}{(1+x)^2}, \qquad y(0) = y(1) = 0.$$

Use the option **Edit/inspect ODE** to confirm that this is the case and then inspect the other options which are available at this point of the program. The default problem can be examined using the prompt **Option: display problem** and you can exit from this part of the program using the prompt **Option: quit**. After quitting the **Edit/inspect ODE** the prompt will become **Method: Collocation**. When this is the case press RETURN and then accept the option **Edit/inspect basis**. There are several sets of basis functions which are available but for the moment when the prompts shows **Select basis: SIN(n*PI*x)** press RETURN and then set the number of basis functions to three. Use the prompt **Option: plot** to examine the default basis which is supplied. When you quit from this option the program will display the prompt **Points data: quit**. Press the SPACE bar to see the various ways in which the points can be specified. For the moment let us use the same points as in example 6.3.3, i.e. 0.25, 0.5 and 0.75. These points can be input using the keyboard or using the option **Points data: standard grid**. Then use the prompt **Points data: quit** in order to compute the approximate solution using the prompt **Option: Calculate solution**. The progress towards finding the solution is indicated displaying the elements of the matrices **A** and **b** as they are found. The final stage is to plot the solution using the prompt **Option: plot solution**. The result should be as in figure 6.3.1.

Exercise 6.3.3 (Computational) Use the program **FEM** to determine an approximate solution of the problem in exercise 6.3.2 using the basis $\{\phi_i(x) = \sin i\pi x\}$, $i = 1, 2, 3, 4$. Compare the results with those in exercise 6.3.2. Try different sets of collocation points and determine the best possible choice.

Exercise 6.3.4 (Computational) Load the program **FEM** and the option **Edit/inspect basis** to input the basis functions $\{\phi_i(x) = x^i(1-x)\}$, $i = 1, 2, 3, 4$. (You will also need to provide ϕ_i' and ϕ_i''.) Use the method of collocation to determine approximate solutions of the default problem. Suggest alternative basis functions and compare the results which are obtained.

In general the matrix associated with the collocation method, (6.3.3), will be full whereas with the finite difference method it is only a tridiagonal matrix. It is possible to obtain basis functions which give a banded matrix. As an example, consider the function

$$B(x) = \begin{cases} 0, & x \le -2, \\ \frac{1}{4}[(2+x)^3], & -2 < x \le -1, \\ \frac{1}{4}[(2+x)^3 - 4(1+x)^3], & -1 < x \le 0, \\ \frac{1}{4}[(2-x)^3 - 4(1-x)^3], & 0 < x \le 1, \\ \frac{1}{4}(2-x)^3, & 1 < x \le 2, \\ 0, & x > 2. \end{cases} \tag{6.3.4}$$

This function is called a *cubic B-spline* since it is a piecewise cubic which has continuous first and second derivatives. It is shown in figure 6.3.2.

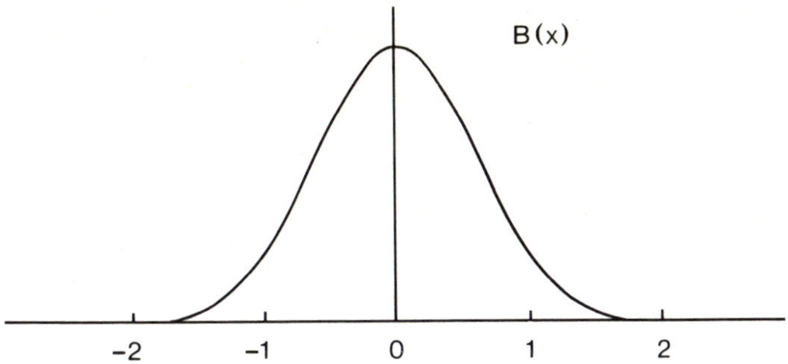

Figure 6.3.2 A cubic B-spline centred about $x = 0$.

In order to construct a set of basis function $\phi_i(x)$, $i = 1, \ldots, N$, we begin by partitioning the interval $(0, 1)$ by a set of N equally spaced points, $0 < x_i = ih < 1$, $(N+1)h = 1$. For each point we define the basis function $B_i(x) = B((x - x_i)/h)$. Notice that outside the interval (x_{i-2}, x_{i+2}) the function $B_i(x)$ is identically zero. In order to apply the collocation method we need to evaluate the B-splines, and their first two derivatives, at the points x_i, $i = 1, \ldots, N$, but this turns out to be particularly easy as shown in table 6.3.1. However, as each B-spline involves five points it is necessary to modify the basis functions near $x = 0$ and $x = 1$. The basis function with modifications are

$$\phi_i(x) = \begin{cases} B_1(x) - \frac{1}{4}B_0(x), & i = 1, \\ B_i(x), & 1 \leq i \leq N-1, \\ B_N(x) - \frac{1}{4}B_{N+1}(x), & i = N, \end{cases} \qquad (6.3.5)$$

to ensure that the boundary conditions are satisfied. Substituting these expressions into equation (6.3.3) we obtain the approximate solution

$$y_N(x) = \sum_{i=1}^{i=N} c_i\phi_i(x), \qquad (6.3.6)$$

where

$$\mathbf{A}\mathbf{c} = \mathbf{b},$$

with $\mathbf{c}^T = [c_1, \ldots, c_M]^T$, $\mathbf{b}^T = [f(x_1), \ldots, f(x_M)]^T$. Only the three main diagonals of the matrix $\mathbf{A}$ are non-zero and are given by

$$\begin{aligned}
a_{1,1} &= -\frac{27}{2h^2}p(x_1) + \frac{3}{4h}p'(x_1) + \frac{15}{4}q(x_1), \\
a_{2,1} &= \frac{6}{h^2}p(x_2) - \frac{3}{h}p'(x_2) + q(x_2), \\
a_{j,j} &= -\frac{3}{h^2}p(x_j) + q(x_j), \qquad j = 2, \ldots, M-1, \\
a_{j,j\pm1} &= \frac{3}{2h^2}p(x_j) \pm \frac{3}{4h}p'(x_j) + \frac{q(x_j)}{4}, \qquad j = 2, \ldots, M-1, \\
a_{M,M} &= -\frac{27}{2h^2}p(x_M) - \frac{3}{4h}p'(x_M) + \frac{15}{4}q(x_M), \\
a_{M-1,M} &= \frac{6}{h^2}p(x_{M-1}) + \frac{3}{h}p'(x_{M-1}) + q(x_{M-1}).
\end{aligned}$$

Exercise 6.3.5 (Computational) Load the program **FEM** to find an approximate solution of the differential equation

$$(y')' - 2\frac{y}{(1+x)^2} = \frac{(2x-4)}{(1+x)^2},$$

subject to the boundary conditions $y(0) = 0$, $y(1) = 0$, using cubic B-splines given by (6.3.4) and five equally spaced collocation points.

Table 6.3.1 Values of B_j, B_j' and B_j'' at $x_j, x_{j\pm1}$.

	x_{j-1}	x_j	x_{j+1}
$B_j(x)$	$1/4$	1	$1/4$
$B_j'(x)$	$3/4h$	0	$-3/4h$
$B_j''(x)$	$3/2h^2$	$-3/h^2$	$3/2h^2$

To do this accept the options **Edit/inspect basis?** and **Select basis: cubic B-splines**. Set the number of splines to five and then accept the prompt **Option: plot**. Now plot the five splines which form the basis. This should produce a display like figure 6.3.3. Notice that the basis functions at either end are modified so that they include only 0, 1 and the collocation points. When you **quit** from this option the program will display the prompt **Option: Calculate solution**; press RETURN and the solution will then be determined. The computed solution can now be plotted using **Option: plot solution**. Compare the solution with that obtained in exercise 6.3.3. What is the minimum number of splines to get a reasonable approximation for the solution? Why are at least three splines required?

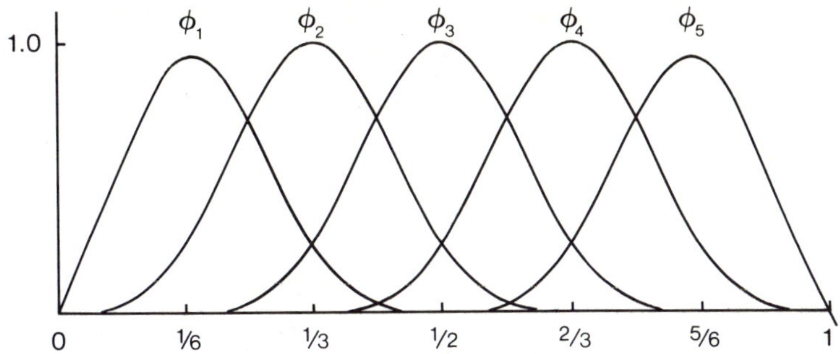

Figure 6.3.3 The five basis functions for exercise 6.3.5.

The choice of the collocation points, x_k, $k = 1, \ldots, N$, is as important as the choice of basis functions. We have suggested that they be evenly spaced between 0 and 1 but this is by no means essential. Other possibilities include the Gaussian quadrature points or the roots of a modified Chebyshev polynomial of degree N.

Exercise 6.3.6 (Computational) Load the program **FEM** with the default problem and use the method of collocation with the basis functions $\{\sin(n\pi x)\}$, $n = 1, 2, 3, 4$ and collocation points as the roots of $T_4(x)$. Compare the result with that obtained using equally spaced points. Why is the solution using the roots of T_4 so poor?

> 6.3.2 Galerkin methods

We now look at an alternative way of minimizing the error between the trial solution Y_N and the exact solution $y(x)$. Let Ω_N be the set of functions which are generated by all possible linear combinations of a given set of basis functions, $\phi_i(x)$, $i = 1, \ldots, N$. Therefore, a function Y_N is a member of Ω_N if and only if

$$Y_N(x) = \sum_{i=1}^{i=N} c_i \phi_i(x). \tag{6.3.7}$$

For each Y_N we can determine the residual

$$r = (pY_N')' + qY_N - f. \tag{6.3.8}$$

In general r will not be a member of Ω_N; however, there will be a projection of r onto Ω_N. Figure 6.3.4 illustrates this situation for $N = 2$.

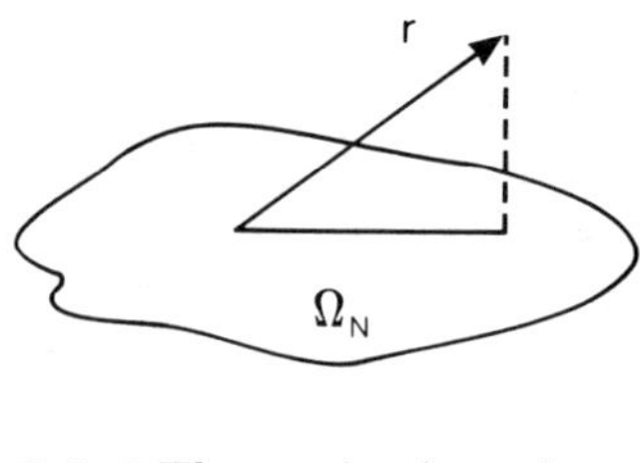

Figure 6.3.4 The projection of r onto Ω_N.

The projection of r onto Ω_N will depend on the coefficients c_i in equation (6.3.7) but it will be zero if they are selected so that r is orthogonal to each basis element. This is the basis of the Galerkin method. We select the coefficients of equation (6.3.7) so that

$$\int_0^1 [(pY_N')' + qY_N - f]\,\phi_k(x)\,dx = 0, \qquad k = 1, \ldots, n. \tag{6.3.9}$$

Substituting for Y_N from equation (6.3.7) and interchanging the sum and the integral produces a system of N linear equations given by

$$\mathbf{A}\mathbf{c} = \mathbf{b},$$

where $b_k = \int_0^1 f(x)\phi_k(x)\,dx$, $k = 1, \ldots, n$, and

$$a_{ki} = \int_0^1 [(p(x)\phi_i')' + q(x)\phi_i]\phi_k\,dx.$$

Integrating by parts reduces this expression to

$$a_{ki} = -\int_0^1 p(x)\phi_k'\phi_i' dx + \int_0^1 q(x)\phi_k\phi_i dx. \tag{6.3.10}$$

Notice that the matrix $\mathbf{A}$ is symmetric, however, only if the basis functions are particularly simple is it possible to express its entries in a closed form. In many cases it will be necessary to resort to numerical integration to find the elements of $\mathbf{A}$. Furthermore, in most cases the matrix $\mathbf{A}$ will be full, but there do exist functions for which it reduces to a tridiagonal form.

Example 6.3.4
Given a set of points $x_j = jh$, $j = 0, \ldots, N$, $Nh = 1$, the functions

$$\phi_j = \begin{cases} \frac{1}{h}(x - x_{j-1}), & x_{j-1} \le x \le x_j, \\ \frac{1}{h}(x_{j+1} - x), & x_j \le x \le x_{j+1}, \\ 0, & \text{otherwise}, \end{cases} \tag{6.3.11}$$

are called linear B-splines and produce a tridiagonal system of linear equations.

Exercise 6.3.7 (Analytical) Show that if the linear B-splines given in Example 6.3.5 are applied to the differential equation $y'' = f(x)$, subject to the boundary conditions $y(0) = y(1) = 0$, then the matrix $\mathbf{A}$ given in equation (6.3.10) reduces to

$$\mathbf{A} = \frac{1}{h}\begin{pmatrix} -2 & 1 & & & \\ 1 & -2 & 1 & & \\ & \ddots & \ddots & \ddots & \\ & & 1 & -2 & 1 \\ & & & 1 & -2 \end{pmatrix}.$$

Exercise 6.3.3 shows the similarity between the Galerkin method and finite difference methods (see *A Simple Introduction to Numerical Analysis* chapter 5). It is possible to show that the difference between the computed solution and the exact solution is proportional to h^2 provided the solution has sufficient bounded derivatives. If cubic splines are used instead then the Galerkin method is fourth order but the linear equations which have to be solved have five elements in most of the rows.

Exercise 6.3.8 (Computational) Load the program **FEM** and the problem

$$(xy')' + x^3 y = -1 + x - x^4, \qquad y(0) = y(1) = 0.$$

Using linear B-splines compute solutions for four basis functions for the option **Galerkin**. What happens as the number of basis functions increases? Repeat this exercise with cubic B-splines.

Exercise 6.3.9 (Computational) Load the program **FEM** with the data file **FE** which contains the problem

$$y'' - 16\pi^2 y = -41\pi^2 \sin(5\pi x)$$

subject to homogeneous boundary conditions. Compute solutions for this problem using the options **Method: Collocation** and **Method: Galerkin** using the basis functions $\{\sin(\pi x), \sin(3\pi x), \sin(5\pi x)\}$. Compare the computed solution with that obtained using (i) cubic B-splines and (ii) linear B-splines. Next the basis functions $\{\sin(n\pi x)\}$, $n = 1, 2, 3$ and suitable collocation points. Now add $\sin(4\pi x)$ to the basis and recompute the solution. Finally, add $\sin(5\pi x)$ and repeat this procedure. What is significant for $n \geq 5$? Why is it better to use the basis $\{\sin(n\pi x)\}$ with $n = 1, 3, 5$ rather than $n = 1, \ldots, 5$?

> 6.3.3 Rayleigh–Ritz methods

Consider the differential equation

$$(p(x)y')' + q(x)y = f(x), \tag{6.3.12}$$

subject to $y(0) = y(1) = 0$. It is possible to show that the solution of this problem is identical with the function which minimizes the functional

$$I(y) = \int_0^1 [p(x)(y')^2 - q(x)y^2 + 2fy]\, dx. \tag{6.3.13}$$

This is called a ***variational principle*** (see Elsgolc (1963), Jordan and Smith (1977)). In general the functional

$$I(y) = \int_0^1 F(x, y, y')\, dx$$

has stationary points which satisfy

$$F_y - \frac{d}{dx}F_{y'} = 0.$$

Exercise 6.3.10 (Analytical) Show that the stationary values of the functional (6.3.13) are equivalent to the solutions of the differential equation (6.3.12). (Hint: look at $I(y+\delta y) - I(y)$.)

Therefore, in order to solve the two-point boundary value problem given by (6.3.12) we need to determine a function which produces the smallest value for the functional I. However, this is as difficult as the original problem and so we shall only attempt to look amongst a subset of all possible functions. For example, given any set of basis functions $\phi_i(x)$, $i = 1, \ldots, N$ we could consider only functions of the form

$$Y_N(x) = \sum_{i=1}^{i=N} c_i \phi_i(x)$$

and select the coefficients c_i to minimize the functional. To do this we substitute Y_N into (6.3.12) and then

$$\begin{aligned} I(c_1, \ldots, c_N) &= \int_0^1 [p(x)(Y_N')^2 - q(x)(Y_N)^2 - 2f(x)Y_N]\, dx \\ &= \int_0^1 \left[p(x)\left(\sum_{i=1}^{i=N} c_i\phi_i'(x)\right)^2 - q(x)\left(\sum_{i=1}^{i=N} c_i\phi_i'(x)\right)^2 \right. \\ &\qquad \left. -2f(x)\sum_{i=1}^{i=N} c_i\phi_i'(x)\right] dx. \end{aligned}$$

Differentiating $I(c_1, \ldots, c_N)$ with respect to c_k, $k = 1, \ldots, N$, will produce a system of linear equations in the unknown coefficients. These equations can then be solved using Gaussian elimination as before. It is possible to show that if the same basis functions are used for the Galerkin and Rayleigh–Ritz methods then the solutions are identical. (See Strang and Fix (1973).)

Exercise 6.3.11 (Analytical) Show that if the linear B-splines given in example 6.3.4 are used in a variational approach to solve the differential equation $y'' = f(x)$, subject to $y(0) = y(1) = 0$, then the system of linear equations which results is identical to that in exercise 6.3.7.

> 6.4 Non-linear problems

It is easy to extend the idea of finite element methods to non-linear equations, but the solution of the resulting non-linear equations for the coefficients is far from simple. For example, consider the differential equation

$$y'' = f(x, y, y'), \qquad y(0) = y(1) = 0. \tag{6.4.1}$$

Now consider a trial function of the form $Y_N = \sum_{i=1}^{i=N} c_i\,\phi_i(x)$ which we substitute into equation (6.4.1) to find

$$\sum_{i=1}^{i=N} c_i \phi_i''(x) = f\left(x, \sum_{i=1}^{i=N} c_i\,\phi_i, \sum_{i=1}^{i=N} c_i \phi_i'\right).$$

If Y_N were the exact solution then this expression would be satisfied identically but we only have N parameters at our disposal. We can now use collocation to ensure that the above equation is satisfied at the points x_k, $k = 1, \ldots, N$. This results in a system of N non-linear equations for the coefficients c_i. For the Galerkin method we ensure that the residual is orthogonal to each of the basis functions, i.e.

$$\int_0^1 [Y_N'' - f(x, Y_N, Y_N'')]\,\phi_k\,dx = 0.$$

Integrating the first term by parts and applying the boundary conditions to each basis function we find

$$-\sum_{i=1}^{i=N}\left(\int_0^1 \phi_i'\phi_i'\,dx\right) c_i = \int_0^1 f\left(x, \sum_{i=1}^{i=N} c_i\,\phi_i, \sum_{i=1}^{i=N} c_i\,\phi_i'\right)\phi_k\,dx,$$

for $k = 1, \ldots, N$. If the basis functions are splines then it is possible to reduce the problem but in general it is beyond the scope of this book. Further details can be found in Strang and Fix (1973).

> Appendix A

> Mathematical Background

In this appendix we shall list some of the background mathematical material which has been used in the text.

> A.1 Taylor's series

Suppose that the power series

$$\sum_{n=0}^{n=\infty} a_n(x-a)^n$$

converges in some interval, $-R < x - a < R$, then the sum of the series has a value for each value of x and so defines a function. Therefore, we may write

$$f(x) = a_0 + a_1(x-a) + a_2(x-a)^2 + \dots, \qquad a - R < x < a + R.$$

If we set $x = a$ it is clear that $a_0 = f(a)$. If we now differentiate with respect to x and set $x = a$, then $a_1 = f'(a)$. Repeating this procedure gives

$$a_n = \frac{f^{(n)}(a)}{n!}.$$

It is possible to show that provided the series converges in some interval then these steps are legitimate and so

$$f(x) = \sum_{n=0}^{n=\infty} \frac{f^{(n)}(a)}{n!}(x-a)^n. \tag{A.1.1}$$

This expression is called the Taylor series expansion of the function f about the point $x = a$.

> A.2 Taylor's series with remainder

The infinite Taylor series expansion (A.1.1) is difficult to manipulate and, furthermore, the function f may not have an infinite number of derivatives. To overcome these difficulties we may use the following result.

Theorem. Taylor's series with remainder.

Suppose that f, f', f'', ..., f^n, and $f^{(n+1)}$ are all continuous in some interval containing $x = a$ and $x = b$; then there is a number θ between a and b such that

$$f(b) = f(a) + f'(a)(b-a) + \tfrac{f''(a)}{2}(b-a)^2 + \tfrac{f'''(a)}{3!}(b-a)^3 + \ldots + \tfrac{f^{(n)}(a)}{n!}(x-a)^n + R_n.$$

where the remainder term R_n is given by

$$R_n = \frac{(b-a)^{n+1} f^{(n+1)}(\theta)}{(n+1)!}.$$

Notice that R_n depends on both a and b but once they are determined θ is fixed.

> A.3 The mean value theorem

Consider a Taylor's series expansion with $n = 1$ and $b = x$ then

$$f(x) = f(a) + (x-a)f'(\theta)$$

or

$$f(x) - f(a) = (x-a)f'(\theta)$$

which is called the *mean value theorem*.

> A.4 Rolle's theorem

Let f be a continuous function in an interval (a, b) such that $f(a) = 0$ and $f(b) = 0$; then there exists a value θ between a and b such that $f'(\theta) = 0$.

Proof. From the mean value theorem we have that

$$f(b) - f(a) = (b - a)f'(\theta),$$

where θ lies between a and b and since $f(a) = 0$ and $f(b) = 0$ we have that $f'(\theta) = 0$ as required.

> A.5 Mean value theorem for integrals

Let f and g be continuous functions over an interval (a, b) and assume that $g(x)$ is positive; then there exists a value θ between a and b such that

$$\int_a^b f(x)\, g(x)\, dx = f(\theta) \int_a^b g(x)\, dx$$

Proof. As g is a continuous function between $x = a$ and $x = b$ there exist constants m and M such that

$$m \leq g(x) \leq M.$$

Accordingly,

$$mf(x) \leq g(x)f(x) \leq Mf(x)$$

and

$$m \int_a^b f(x)\, dx \leq \int_a^b g(x)f(x)\, dx \leq M \int_a^b f(x)\, dx.$$

Therefore, there exists a value $\mathcal{M}$ between m and M such that

$$\int_a^b f(x)g(x)\, dx = \mathcal{M} \int_a^b f(x)\, dx.$$

Furthermore, since g is continuous there is a value θ such that $g(\theta) = \mathcal{M}$ which concludes the proof.

> Appendix B

> Finite Difference Approximations

In order to construct finite difference approximations for derivatives we consider the Taylor series expansion of a function $y(x)$ about $x = a$, for example,

$$y(a+h) = y(a) + hy'(a) + \frac{h^2}{2!}y''(\theta).$$

We now re-arrange this expression to give

$$\frac{y(a+h) - y(a)}{h} = y'(a) + \frac{h}{2}y''(\theta)$$

and so we can use the expression

$$\frac{y(a+h) - y(a)}{h}$$

as an approximation for $y'(a)$. Furthermore, the error in such an approximation is proportional to h provided the function y has a bounded derivative. We say that this approximation is $O(h)$. Alternatively, we can consider the Taylor series expansions

$$y(a+h) = y(a) + hy'(a) + \frac{h^2}{2!}y''(a) + \frac{h^3}{3!}y'''(a) + \ldots$$

and

$$y(a-h) = y(a) - hy'(a) + \frac{h^2}{2!}y''(a) - \frac{h^3}{3!}y'''(a) + \ldots.$$

Subtracting, the second expansion from the first we have

$$y(a+h) - y(a-h) = 2hy'(a) + 2\frac{h^3}{3!}y'''(a) + \ldots$$

and so

$$\frac{y(a+h) - y(a-h)}{2h}$$

may be used to approximate $y'(a)$ with error $O(h^2)$.

We can extend the above approach to construct approximations for higher derivatives as follows:

$$y(a+h) = y(a) + hy'(a) + \frac{h^2}{2!}y''(a) + \frac{h^3}{3!}y''' + \frac{h^4}{4!}y^{iv} + \ldots$$

and

$$y(a-h) = y(a) - hy'(a) + \frac{h^2}{2!}y''(a) - \frac{h^3}{3!}y''' + \frac{h^4}{4!}y^{iv} + \ldots.$$

Adding these two expressions together we obtain

$$y(a+h) + y(a-h) = 2y(a) + h^2y''(a) + \frac{h^4}{12}y^{iv} + \ldots$$

from which

$$\frac{y(a+h) - 2y(a) + y(a-h)}{h^2}$$

is an approximation for $y''(a)$ with error $O(h^2)$.

Bibliography

[1] Bender C M and Orszag S A 1978 *Advanced Methods for Scientists and Engineers* (New York: McGraw Hill)

[2] Burden R L, Faires J D and Reynolds A C 1981 *Numerical Analysis* 2nd edn, ed. Prindle, Weber and Schmidt (Belmont, CA: Wadsworth)

[3] Cheney E W 1966 *Introduction to Approximation Theory* (New York: McGraw Hill)

[4] Cheney E W and Kincaid D 1980 *Numerical Mathematics and Compution* (Monterey, CA: Brook/Cole)(Belmont, CA: Wadsworth)

[5] Churchill R V 1963 *Fourier Series and Boundary Value Problems* International Student Edn (New York: McGraw Hill)

[6] Clenshaw C W and Curtis A R 1960 A method for numerical integration on an automatic computer *Num. Math.* **2** 197–205

[7] Coddington E A and Levinson N 1955 *Theory of Ordinary Differential Equations* (New York: McGraw Hill)

[8] Davis P J and Rabinowitz P 1975 *Methods of Numerical Integration* (New York: Academic)

[9] de la Vallee Poussin C J 1919 *Leçon sur l'Approximation des Fonctions d'une Variable Réelle* (Paris: Gauthier-Villars) Reprinted 1952

[10] Elsgolc L E 1963 *Calculus of Variations* (Oxford: Pergamon)

[11] Fröberg C E 1985 *Numerical Mathematics* (New York: Benjamin/Cummings)

[12] Fox L 1962 *The Numerical Solution of Ordinary and Partial Differential Equations* (Oxford: Pergamon)

[13] Fox L and Mayers D F 1965 *Computing methods for Scientists and Engineers* (Oxford: Oxford University Press)

[14] Gerald C F 1977 *Applied Numerical Analysis* 2nd edn (New York: Addison–Wesley)

[15] Gerald C F and Wheatley P D 1984 *Applied Numerical Analysis* 3rd edn (New York: Addison–Wesley)

[16] Harding R D 1982 *Graphs and Charts on the BBC Microcomputer* (London: Acornsoft)

[17] Harding R D 1986 *A Mathematical Toolkit: Numerical Routines with Applications in Engineering, Mathematics and the Sciences* (Bristol: Adam Hilger)

[18] Harding R D and Quinney D A 1986 *A Simple Introduction to Numerical Analysis* (Bristol: Adam Hilger)

[19] Isaacson E and Keller H B 1966 *Analysis of Numerical Methods* (New York: Wiley)

[20] Jordan D W and Smith P 1977 *Non-linear Ordinary Differential Equations* Oxford Applied Mathematics and Computing Series (Oxford: Oxford University Press)

[21] Maehly H J 1960 Methods for fitting rational approximations *ACM J.* **7** 150–62

[22] Maehly H J 1963 Methods for fitting rational approximations *ACM J.* **10** 257–77

[23] Padé 1892 Sur la représentation approchée d'une fonction par des fractions rationelles *Anales Sci. l'École Normales Supérieure* **9** supplement 1–93

[24] Philips G M and Taylor P J 1973 *Theory and Application of Numerical Analysis* (New York: Academic)

[25] Remes E Ya 1935 *On the best Approximation of Functions in the Chebyshev Sense* (Kiev: Ukrainian)

[26] Sawyer W W 1977 *A First Look at Numerical Functional Analysis* Oxford Applied Mathematics and Computing Series (Oxford: Oxford University Press)

[27] Strang G and Fix J F 1973 *An Analysis of the Finite Eelement Method* Prentice–Hall Series in Automatic Computation (New York: Prentice Hall)

[28] Stoer J and Burlisch R 1980 *Introduction to Numerical Analysis* (Berlin: Springer)

[29] Williams P W 1973 *Numerical Computation* (Camden, NJ: Nelson)

Index